Defending Einstein

Hans Reichenbach, a philosopher of science who was one of five students in Einstein's first seminar on the general theory of relativity, became Einstein's bulldog, defending the theory against criticism from philosophers, physicists, and popular commentators. This book chronicles the development of Reichenbach's reconstruction of Einstein's theory in a way that clearly sets out all of its philosophical commitments and physical predictions as well as the battles that Reichenbach fought on its behalf, in both the academic and popular press. The essays include reviews and responses to philosophical colleagues, such as Moritz Schlick and Hugo Dingler; polemical discussions with the physicists Max Born and D. C. Miller; and popular articles meant to clarify aspects of Einstein's theories and set out their philosophical ramifications for the layperson. This book is a window into the development of scientific philosophy and the role of the philosopher at a time when physics and philosophy were both undergoing revolutionary changes in content and method.

Steven Gimbel is associate professor of philosophy at Gettysburg College in Pennsylvania, where he was named Luther W. and Bernice L. Thompson Distinguished Teacher in 2005. He has contributed to *Philosophy of Science, British Journal of Philosophy of Science*, and *Studies in History and Philosophy of Modern Physics.*

Anke Walz is assistant professor of mathematics at Kutztown University. Her work on the bellows conjecture relating to flexible polyhedrals with R. Connely and I. Sabitov has appeared in *Beiträge zur Algebra und Geometrie* and has received coverage in *Science* and *Scientific American.*

Defending Einstein

Hans Reichenbach's Writings on Space, Time, and Motion

Edited by

Steven Gimbel

Gettysburg College

Anke Walz

Kutztown University

CAMBRIDGE UNIVERSITY PRESS
Cambridge, New York, Melbourne, Madrid, Cape Town,
Singapore, São Paulo, Delhi, Tokyo, Mexico City

Cambridge University Press
32 Avenue of the Americas, New York, NY 10013-2473, USA

www.cambridge.org
Information on this title: www.cambridge.org/9780521371162

First published 2006
First paperback edition 2011

A catalog record for this publication is available from the British Library

Library of Congress Cataloging in Publication data

Reichenbach, Hans, 1891–1953.
Defending Einstein : Hans Reichenbach's writings on space, time, and motion /
edited by Steven Gimbel, Anke Walz.
p. cm.
ISBN-13: 978-0-521-85958-5 (hardback)
ISBN-10: 0-521-85958-1 (hardback)
1. Relativity (Physics) 2. Relativity (Physics) – Philosophy. 3. Physics –
Philosophy. I. Gimbel, Steven, 1968– II. Walz, Anke. III. Title.
QC173.55.R439 2006
530.11 – dc22 2005036291

ISBN 978-0-521-85958-5 Hardback
ISBN 978-0-521-37116-2 Paperback

Contents

Introduction

Fifty years after the untimely death of Hans Reichenbach, the evolution of his views on space, time, and motion is receiving much due critical attention. While his mature views, expressed most famously in his book *Philosophy of Space and Time*, have long been commented upon, challenged, and amended by some of the most important figures in contemporary discussions of the philosophical foundations of physics, only in the last decade or so – a couple of generations removed from the heyday of logical empiricism – is the work of Reichenbach and contemporaries like Moritz Schlick and Rudolf Carnap being widely considered in terms of its value to the history of ideas.

This collection seeks to contribute to that growing conversation by bringing together English translations of nine essays from 1920–25, the period preceding *Philosophy of Space and Time*, that have not appeared in earlier collections of Reichenbach's writings. These articles range from technical discussions published in scientific journals, to overtly philosophical discussions and responses to philosophical opponents published in philosophical journals, to semipopular pieces designed to set out Reichenbach's interpretation of relativity theory in clear, explicit terms. It is hoped that by providing access to additional "data points," the discussion of the emergence of Reichenbach's later views may be advanced.

The first half of the 1920s was a period of crucial importance both to the burgeoning movement that would become analytic philosophy and to the philosophical development of Hans Reichenbach personally. This was the period that saw Schlick move to Vienna, the publication of Ludwig Wittgenstein's *Tractatus Logico-Philosophicus*, and Alfred Tarski's initial work on set theory. It was also the time in Reichenbach's life when he sought to put his name on the philosophical map. Leaving an engineering position in Berlin, he accepted his first academic post, *Privatdozent* at the Technische Hochschule in Stuttgart. As an assistant to the physicist Erich Regener, he undertook a diverse teaching load covering courses in modern physics, radio technique, and philosophy; his research largely focused upon developing a rigorous epistemic foundation for the theory of relativity – a project that he had started in Berlin after attending Einstein's lectures at the university there and that had led to the publication of his first book, *Theory of Relativity and A Priori Knowledge*, earlier in 1920.

This rigorous epistemic foundation would be laid out in terms of what Reichenbach called a "constructive axiomatization," appearing in full for the special theory of relativity in Reichenbach's 1924 publication *Axiomatization of the Theory of Relativity*. This axiomatization was set out with two equally ambitious goals in mind. First, it was to give a complete and unambiguous account of all of the empirical and definitional commitments of Einstein's theories. This would allow for an objective assessment of the strength of experimental evidence in favor of the theory, as well as a determination of exactly what portions of the theory remained empirically unconfirmed, and could be used to clarify and comment reasonably upon objections raised to the theory. This final concern was no small matter as Reichenbach notes that during this period, in discussions of the theory of relativity "dogmatic understanding is found next to clear insight."[1]

Armed with his analysis of relativity theory and his ability to translate complex technical notions into plain language, especially through the use of colorful analogies, Reichenbach set out to be Einstein's bulldog. Reichenbach would confront Einstein's critics on their own turf, whether in the scholarly arena or the popular press, whether the objection was based upon scientific, epistemological, logical, or intuitive grounds. He took it as his mission to defend and popularize the relativity theory, writing articles and responses in professional physics and philosophy journals, books for the general reader, and articles in popular science magazines, even giving a series of radio broadcasts.

Reichenbach saw the discussion of Einstein's work on relativity advancing on two distinct fronts. On one side were those commentators who took issue with the theory of relativity on physical or a priori grounds but who had little or no understanding of the details of the theory, and on the other side stood those opposed to Einstein despite being well schooled in physical theory. The first group contained its share of cranks. For example, in "Einstein's Theory of Motion" (Chapter 6 in this collection), Reichenbach recounts tales of such intellectual "off-key notes," including one opponent who wrote to the Swedish Academy demanding the Nobel Prize for uncovering "the 'miscalculation' in Einstein's theory." Reichenbach never held back on his sharp wit in dealing

[1] Reichenbach 1920*a*, p. 2.

with such attacks, but far from engaging in mere ad hominem argument, he sought to clearly refute all objections.

But the group also contained a fair number of respected philosophers. Indeed, Reichenbach refers to this period as "a time of increasing philosophical unrefinement (including among the tenured philosophers)."[2] Criticism of relativity came from virtually all corners of the philosophical landscape, most notably from Neo-Kantians and Machian Positivists. A core group of Neo-Kantians, such as Ewald Sellien and Ilsa Schneider, realized the threat that relativity, with its lack of absolute time and its use of non-Euclidean geometry, posed to an orthodox understanding of the transcendental aesthetic and sought to provide a priori grounds for the theory's failure. To this audience, Reichenbach could point to his *Theory of Relativity and A Priori Knowledge*, in which he sought to salvage what he thought important in Kant and jettison that which he saw as problematic.

Not all Neo-Kantians, of course, took such a dogmatic and flawed postion. Ernst Cassirer, in particular, is singled out by Reichenbach as having "awakened Neo-Kantianism from its 'dogmatic slumber', while its other adherents carefully tried to shield it from any disturbance by the theory of relativity."[3] Having attended Cassirer's lectures in 1913–14, and having been exposed to Kant earlier by Alois Riehl, Ernst von Aster, and Karl Stumpf, Reichenbach's work maintained a strong Kantian element throughout the entire span of his writings, but it is especially evident in this early period.

Beside the Neo-Kantian discussions, Machian positivists were also quite vocal on issues pertaining to the theory of relativity.

[2] Reichenbach 1920*b*, p. 1.
[3] Reichenbach 1921*a*, p. 25.

Ernst Mach himself simultaneously played the conflicting roles of forefather and opponent of the theory of relativity. Reichenbach's treatment of Mach's work is thereby both deferential and critical. We find repeated discussions of the insights that led Mach to overthrow the needlessly metaphysical foundation upon which Newton grounded his mechanics, but we also see repeated criticisms of Mach's attempt to respond to physical aspects of this question with a priori argumentation. The complete instantiation of a relativistic mechanics required not only the philosophical shift away from Newton provided by Mach, but also a physical alternative missing in his work. This, of course, Reichenbach finds in the work of Einstein.

Upon Mach's death, his successors divided along this fault line. Joseph Petzold seized on the "forefather of relativity" angle and sought to reconstruct the theory of relativity in purely phenomenal terms,[4] while Hugo Dingler aligned himself with Mach's opposition to Einstein. In his "Reply to H. Dingler's Critique of the Theory of Relativity" (Chapter 3 in this collection), Reichenbach systematically attacks Dingler's stance that seeks to undermine the observer dependence of distance and duration in the theory of relativity by trying to marry Mach's a priori relativistic understanding of translation to a Newtonian absolutist view of rotation.

Among the scientific objections raised to relativity theory, Reichenbach spends the most time considering suggested empirical mechanisms for the absolute determination of simultaneity for nonlocal events, such as the slow transport of clocks (see Chapter 7 in this collection). In other physical discussions he seeks to

[4] For a full account of the Reichenbach/Petzold correspondence, see Hentschel 1991*b*.

clarify the impact on relativity theory of experiments designed to test the principle of the constancy of the speed of light using observations of the eclipses of the moons of Jupiter and empirical challenges to the results of the Michelson experiment (see Chapter 11 in this collection).

But not all commentators of the period were to be considered foes; there were friends involved in the discussion as well. Most notable among these is Moritz Schlick, a former student of Max Planck and the author of *Space and Time in Contemporary Physics*, a book that counted among its admirers Albert Einstein himself. Much has been made of Reichenbach's comments about Schlick's major work, *General Theory of Knowledge*, in Reichenbach's *Theory of Relativity and A Priori Knowledge* and the correspondence between Schlick and Reichenbach in 1920,[5] but less has been said about Reichenbach's published review of the work.[6] This review is included as Chapter 1 in this collection.

More complex is the case of Hermann Weyl, the accomplished mathematical physicist. Weyl not only understood the theory of relativity, but proposed an extension of it representing the first full attempt at a unified field theory. Reichenbach and Weyl discussed the foundations of relativity theory by correspondence, and Reichenbach refers deferentially to Weyl throughout his works, even mentioning Weyl's *Space, Time, Matter* as a preferred introduction to relativity for laymen. But the relationship took a turn for the worse when Weyl published a blisteringly negative review of Reichenbach's *Axiomatization of the Theory of Relativity*, calling the work "unsatisfactory" as a result of being "too cumbersome and overly obscure." Reichenbach responds in kind in "On the

[5] See Friedman 1994 and 1999 and Coffa 1990.
[6] The notable exception is Hentschel 1991*a*.

Physical Consequences of the Axiomatization of the Theory of Relativity" (Chapter 12 in this collection), offering what he terms a "strenuous defense" against Weyl's objections.[7]

In addition to attempting to clarify the muddled state of the discussion of relativity at that time, the second goal of Reichenbach's constructive axiomatization was to serve as a model for future philosophical research. Kant's doctrine of the synthetic a priori had been successful, Reichenbach contended, in demonstrating the role of constitutive elements in the groundwork of our knowledge. However, in light of the advances of mathematics at the end of the nineteenth century and of physics at the start of the twentieth, Kant's epistemology had displayed a fatal flaw. It was unable to free itself from the limitations of the Euclidean and Newtonian theories that it had been constructed to justify. Reichenbach sought to develop a "method of scientific analysis"[8] that contained the constitutive elements of Kant, but that replaced the apodictic nature of these concepts with a sophisticated sense of theory dependence. In this way, individual analyses would be theory-specific, but the epistemological basis of the method of the

[7] For more on Reichenbach and Weyl, see Ryckman 1994 and 1996 and Rynasiewicz 2002 and 2005.

[8] The term "*wissenschaftsanalytische Methode*" has been translated in this collection as the "method of scientific analysis" instead of using the phrase "method of logical analysis" employed by Maria Reichenbach in her translations. The reasons for this choice are twofold: first, it is closer to the literal translation of the term and, second, it distances the analytic process proposed in Reichenbach's version of Logical Empiricism from that of Rudolf Carnap's brand of Logical Positivsm, with which it is often wrongly conflated. Unlike Carnap's attempts to construct empirical claims using only logical and empirical terms, the observational atoms in Reichenbach's axiomatizations are still quite pregnant with theoretical terms, albeit terms of theories past. This difference renders the use of the adjective "scientific" instead of "logical" in this central phrase warranted in this context. For a discussion of the theory-laden aspects of observation sentences in Reichenbach's axiomatic method, see Gimbel 2004.

analyses would no longer be pregnant with the concepts of any particular theory.

In Michael Friedman's terminology, Reichenbach tried to move epistemology from the synthetic a priori to the "relativized *a priori*."[9] But like Kant and Hume before him, whose attempts to radically revise philosophical discourse fell stillborn from the presses, Reichenbach's axiomatization received less attention and acclaim than he had hoped it would, being too technical for philosophical audiences and too philosophical for scientific readers. This work, however, along with the aid of Einstein and Max Planck, was sufficient to land Reichenbach a teaching position in natural philosophy at the University of Berlin in 1926. Returning to Berlin, Reichenbach would go on to publish his best-known work, *Philosophy of Space and Time*, organize the Gesellschaft für Empirische Philosophie, and begin and edit the journal *Erkenntnis* in conjunction with Rudolf Carnap in Vienna and later in Prague.

It has been noted by several significant commentators[10] that Reichenbach's views on space, time, and motion underwent a significant alteration in the years between *Theory of Relativity and A Priori Knowledge* and *Philosophy of Space and Time*. He began the decade eschewing Poincaré's conventionalist doctrine and ended it as a self-proclaimed adherent. Different accounts are offered to explain this shift. With respect to this discussion, three foci of Reichenbach's discussions in the following essays should be introduced: the undermining of Newton's doctrine of absolute time, the constructability of a light geometry, and the emergence of Reichenbach's empiricist version of geometric conventionalism.

[9] See Friedman 1994 and 1999.

[10] For example, see Friedman 1994 and 1999, Coffa 1990, and Ryckman 1994, 1996, and 2003.

While Reichenbach's views on the epistemology of geometry have far and away garnered the most critical attention from scholars, it is not the topic that dominates Reichenbach's own works. While geometric concerns are certainly discussed in the works of this period, much more attention is paid by Reichenbach during this period to issues of time. Indeed, significant discussions of absolute and relativistic time may be found in the majority of the articles in this collection, and time is the sole matter of attention in several of them.

This should not be unexpected. In section II of *Theory of Relativity and A Priori Knowledge*, Reichenbach argues that the main philosophical upshot of Einstein's special theory of relativity is a correction to the classical notion of time. We find this position maintained throughout the articles in this collection, which are largely devoted to mining Reichenbach's axiomatization of the special theory for its philosophical ramifications. Reichenbach argues that among the most significant results of this axiomatic project are (1) the positing of a distinction between the epistemological and physical senses of absolute time, and (2) the explicit determination of what empirical evidence supports Einstein's arguments for the dissolution of absolute time in the physical sense.[11]

But significant attention is also paid to geometrical concerns. Of the initial results of the constructive axiomatization, Reichenbach repeatedly cites the possibility of the construction of a light geometry in Minkowski space-time as one of the most important. Reichenbach argues in his "Report on an Axiomatization of

[11] For a discussion of the introduction of Reichenbach's famous ε-formulation of Einstein's definition of simultaneity and the role it plays in his view of time in the special theory of relativity, see Rynasiewicz 2002.

Einstein's Theory of Space-Time" (Chapter 4 in this collection) that it is possible to determine a space-time geometry univocally without the use of material rods or clocks. The combination of empirical axioms governing the behavior of light and needed coordinative definitions are by themselves sufficient to completely describe the geometric structure of Minkowski space-time. Matter axioms need only specify that material bodies conform to the geometry specified by light signals.

This emphasis on the constructability of a light geometry suited Reichenbach's aims in a couple of ways. For his professional audience, the ability to derive the space-time metric from purely optical phenomena placed the foundation of Einstein's theory on almost entirely firm ground. With only a couple of small exceptions, the experimental evidence for each of the light axioms used to create the light geometry was widely accepted within the scientific community. One could ground the theory's validity upon innocuous results from optics without reference to the more contentious descriptions of the behavior of matter. This was also useful for his discussions geared toward the popular audience, where Reichenbach could now set out the counterintuitive results of the theory as flowing from an almost fully supported and completely independent evidentiary basis.

Unfortunately, Reichenbach's claims of the complete and unique constructability of the light geometry failed. Weyl pointed out to Reichenbach that the geometry of relativistic space-time was, in fact, underdetermined by the purely optical means Reichenbach employed. If one makes certain topological assumptions and restricts consideration to spaces free of singularities, then the light geometry is constructable in the way that Reichenbach requires. But this assumption requires an axiom, an empirical foundation. One can easily solve this problem through the invocation of a matter axiom defining the notion of straight line, but to do so would be to undermine the point of the light geometry. Reichenbach tries to finesse the question in (1926), but ultimately relies on a mere assumption.[12]

Most significant, perhaps, is the emergence of Reichenbach's mature position on geometric conventionalism. In "The Philosophical Significance of the Theory of Relativity" (Chapter 7 in this collection),[13] Reichenbach argues that "the concept of time has already received a decisive correction in the special theory of relativity. The concept of space, on the other hand, will only be subject to modification in the general theory." This modification requires the introduction of Reichenbach's famous notion of universal forces, or "forces of type X" as he refers to them here, which affect all objects in the same way and against which it is impossible to insulate. Reichenbach argues that such forces appear fictitious and are rightfully ignored generally; but in treating gravitation as a space-time curving force, Einstein has required the inclusion of universal forces in the general theory of relativity. While this is surely the introduction of Reichenbach's complete argument for geometric conventionalism, shades of it appear earlier. In his article "Einstein's Theory of Space" (Chapter 2 in this collection), for example, we see elements of this argument from universal forces begin to appear when he considers a D-shaped wire the size of a town and a force that contracts all objects in the same fashion that is active only in the immediate vicinity of the center point of the diameter. Such a force, he argues, would be directly undetectable because it shrinks the measuring rods, but the geometric

12 For a discussion of this point, see Rynasiewicz 2005.

13 This article of 1922 should not be confused with Reichenbach's contribution of the same title in P. A. Schillp's *Albert Einstein: Philosopher-Scientist.*

effects would be determinable in noting the deviation from when we divide the circumference of the arc by the length of the diameter.

The increased opportunity to trace the development of such points in Reichenbach's thought gives us hope that the pieces included in this collection will be useful to contemporary scholars of logical empiricism in discussing the evolution of Hans Reichenbach's thought on space, time, and motion and the development of analytic philosophy more generally.

We would like to thank Robert Rynasiewicz and especially Flavia Padovani for their help with the translations, Russell Harbin, and Paul Gimbel for help with the illustrations. We are especially grateful to Maria Reichenbach for her gracious permission to publish these translations.

References

Coffa, J. Alberto. (1990) *The Semantic Tradition from Kant to Carnap.* Cambridge: Cambridge University Press.

Friedman, Michael. (1994) "Geometry Convention, and the Relativized *A Priori*: Reichenbach, Carnap, and Schlick." In *Logic, Language, and the Structure of Scientific Truth*, ed. Wesley Salmon and Gereon Wolters. Pittsburgh: University of Pittsburgh Press, pp. 21–34.

Friedman, Michael. (1999) *Logical Positivism Reconsidered.* Cambridge: Cambridge University Press.

Gimbel, Steven. (2004) "Unconventional Wisdom: Theory-specificity in Reichenbach's Geometric Conventionalism." *Studies in History and Philosophy of Modern Physics*, vol. 35, pp. 457–81.

Hentschel, Klaus. (1991*a*) "Die vergessene Rezension der 'allgemeinen Erkenntnislehre' Moritz Schlicks durch Hans Reichenbach – Ein Stück Philosophiegeschichte." In *Erkenntnis Orientated: A Centennial Volume for Rudolf Carnap and Hans Reichenbach*, ed. Wolfgang Spohn. Dordrecht: Kluwer, pp. 11–28.

Hentschel, Klaus. (1991*b*) *Die Korrespondenz Petzold – Reichenbach: Zur Entwicklung der "wissenschaftlichen Philosophie" in Berlin.* Berlin: ERS Verlag.

Howard, Don. (1984) "Realism and Conventionalism in Einstein's Philosophy of Science: The Einstein/Schlick Correspondence." *Philosophia Naturalis*, vol. 21, pp. 416–29.

Reichenbach, Hans. (1920*a*) *Theory of Relativity and A Priori Knowledge.* Berkeley: University of California Press, 1965.

Reichenbach, Hans. (1920*b*) Review of Moritz Schlick's *Algemeine Erkenntnislehre. Zeitschrift für angewandte Psychologie*, vol. 16, pp. 341–3.

Reichenbach, Hans. (1920*c*) "*Die Einsteinsche Raumlehre.*" *Die Umschau*, vol. 24, no. 3, pp. 402–5.

Reichenbach, Hans. (1921*a*) "The Present State of the Discussion on Relativity." In *Modern Philosophy of Science*, ed. Maria Reichenbach. New York: Routledge and Kegan Paul, 1959, pp. 1–45.

Reichenbach, Hans. (1921*b*) "*Erwiderung auf H. Dinglers Kritik an der Relativitätstheorie.*" *Physikalische Zeitschrift*, vol. 21, pp. 379–84.

Reichenbach, Hans. (1921*c*) "*Bericht über eine Axiomatik der Einsteinschen Raum-Zeit-Lehre.*" *Physikalische Zeitschrift*, vol. 22, pp. 683–7.

Reichenbach, Hans. (1921*d*) "*Die Einsteinsche Bewegungslehre.*" *Die Umschau*, vol. 27, pp. 501–5.

Reichenbach, Hans. (1922*a*) "*Relativitätstheorie und absolute Transportzeit.*" *Zeitschrift für Physik*, vol. 9, pp. 111–17.

Reichenbach, Hans. (1922*b*) "*La signification philosophique de la théorie de la relativité.*" *Revue Philosophique de la France et de l'Étranger*, vol. 94, pp. 5–61.

Reichenbach, Hans. (1924*a*) *Axiomatization of the Theory of Relativity.* Berkeley: University of California Press, 1969.

Reichenbach, Hans. (1924*b*) "*Planetenuhr und Einsteinsche Gleichzeitkeit.*" *Zeitschrift für Physik*, vol. 33, no. 8, pp. 628–34.

Reichenbach, Hans. (1925) "*Über die physikalische Konsequenzen der relativischen Axiomatik.*" *Physikalische Zeitschrift*, vol. 34, pp. 32–48.

Reichenbach, Hans. (1926) *Philosophy of Space and Time.* New York: Dover, 1957.

Reichenbach, Hans. (1978) *Hans Reichenbach: Selected Writings, 1909–1953*, ed. Maria Reichenbach and Robert S. Cohen. Dordrect: Reidel.

Ryckman, T. A. (1994) "Weyl, Reichenbach, and the Epistemology of Geometry." *Studies in the History and Philosophy of Science*, vol. 25, pp. 831–70.

Ryckman, T. A. (1996) "Einstein Agonists: Weyl and Reichenbach on Geometry and the General Theory of Relativity." In *Origins of Logical Empiricism*, ed. Ronald Giere. Minneapolis: University of Minnesota Press, pp. 165–209.

Ryckman, T. A. (2003) "Two Roads from Kant: Cassirer, Reichenbach, and General Relativity." In *Logical Empiricism: Historical and Contemporary Perspectives*, ed. Paolo Parrini. Pittsburgh: University of Pittsburgh Press, pp. 159–93.

Rynasiewicz, Robert. (2002) "Reichenbach's ε-Definition of Simultaneity in Historical and Philosophical Perspective." *Vienna Circle Institute Yearbook*, vol. 10, pp. 121–9.

Rynasiewicz, Robert. (2006) "Weyl vs. Reichenbach on Lichtgeometrie." In *The Universe of General Relativity*, ed. J. Eisenstädt and A. J. Kox. Boston: Birkhauser, in press.

1

Review of Moritz Schlick's *General Theory of Knowledge*

Moritz Schlick. General Theory of Knowledge.
Naturwissenschaftliche Monographien und Lehrbücher.
1. Editor: Berliner and Pütter. Berlin. Julius Springer. 1918. 344 pp.
M. 18. C.

[1] The philosophical literature today is moving off in two divergent directions. Both share the goal of understanding the foundational facts underlying the knowledge process and which can only be experienced individually; but while one direction strives to illuminate and clarify these facts by breaking down the terms in which the facts are expressed, it suits the needs of the other to obscure them. The latter, of course, is always possible since the words, which surround the facts like a mushy porridge, are difficult to control and in their emptiness are often completely irrefutable. It should therefore be stated from the start that Schlick's representation belongs to the first direction, and this must be considered its

Translated from *Zeitschrift für Angewandte Psychologie*, vol. 16 (1920), pp. 341–3.

initial virtue in a time of increasing philosophical unrefinement (including among the tenured philosophers).

[2] Its second merit is that it bases its critique upon a systematic construction including a closed framework of concepts, and does not at all shy away from manipulating and shaping them until they are integrated into this framework. This is an act deserving of thanks because it creates a work of thought that gives rise to discussions whose cognitive value can be clearly assessed. In addition, the language used is pleasingly sober, completely excluding all affected attitudes and not confusing ethical sentiment with epistemological clarification. Furthermore, the investigation is grounded in an expert knowledge of the theoretical natural sciences, particularly physics, so that all statements relevant to the exact sciences are unobjectionable. Schlick is one of the first to apply the results of Einstein's theory of relativity to the philosophy of space. However, one reproach from which we will not always be able to excuse Schlick, is that the clarity of his presentation is not always the result of overcoming the problem at hand, but occasionally arises from avoiding what is truly problematic.

[3] Following his scientific approach, Schlick's foundation rests upon the naive juxtaposition of thing and consciousness, of the actual object and its representation in thought. However, he opposes viewing these representations as a repetition of the object, as a shape somehow similar to the thing. "Representation" here only means coordination with an arbitrarily chosen sign, and scientific knowledge signifies nothing more than the coordination of a system of signs to the world of actual things. The surely untenable Scholastic definition of truth – agreement of the representation with its object – is thereby meaningless because such an agreement is completely impossible; we are only able to require

that statements about the same object always agree, and truth for him is thus equivalent to the univocality of the coordination. However, while we do have an intuitive representation of the object, this intuition has nothing to do with knowledge. It is merely the manner in which our sense organs represent an impression, therefore the intuited qualities of each individual sense are quite different, e.g., visual space has a completely different structure than tactile space. But physical space is not intuitive, it is merely a reference system of coordinates in which we arrange the objects. That is why we are not able to posit anything *a priori* about this space, e.g., that it is non-Euclidean; it remains the task of experience to determine which geometry allows for the simplest formulation of natural laws. (For Einstein this is a Riemannian geometry of non-constant curvature.) Hence, space is not a property of real objects; it is the scheme of order of the things we put in it.

[4] This theory of intuition undoubtedly has one essential success; it makes it possible to embed Einstein's theory of gravitation into philosophy. But this theory appears to me to be one of those areas in which Schlick finishes precisely where the real problems first begin. Visual space, tactile space, and physical space are all non-Euclidean. So what space is now properly Euclidean? It cannot be denied that Euclidean space somehow still remains our preferred setting, so what is the source of this compulsion that leads every impartial mind to come to believe in the truth of the parallel axiom even though it is physically invalid? We cannot dismiss it by using the catchphrase "habituation." The question is open as to whether it will be psychology or epistemology that will be able to answer this question. Schlick would assign it to psychology without even a hint at the path to a solution.

[5] If knowledge is implemented through concepts and judgments, the Kantian question that culminated in the demonstration of the synthetic *a priori* re-emerges. Schlick takes on this question very systematically and in great detail. Analytic judgments are easy to account for because they merely purport to repeat what was already known. However, Schlick denies the existence of synthetic *a priori* judgments. That he denies the existence of the intuitively obtained truths (Kant's transcendental aesthetic) follows from his theory of the intuition in which intuition is irrelevant to knowledge; however, his rejection of the purely conceptual (Kant's transcendental analytic) seems to me not justified on the basis of this rationale. Because the Kantian question, "What are the conditions for knowledge?" still remains for him, he subsumes even the problem of causality under it; but once the existence of such propositions preceding knowledge is admitted, their central position is established, and the name is no longer a concern. Instead, questions should be raised for other presuppositions of knowledge beyond causality. It seems as if Schlick, in endeavoring to present the knowledge process as a simple scheme, had to overly distort the facts. Where the analysis of the problem of probability should have led him to new postulates underlying our knowledge, instead it resulted in an unfortunate reduction of probability judgments to a statement of a system of conditions.

[6] Here lies the limitations of Schlick's work. While it is entirely based upon the foundation that recent mathematical and physical analyses have provided for epistemology, he does not contribute to that foundation. To some extent, it gives excellent critiques of other systems (e.g., the critique of positivism has a very pleasing logical structure), and it dwells upon the problem of judgment with organized thoroughness, e.g., in the reduction of all

Aristotelian forms to a single one: the theory of implicit definitions. His fine way of presenting material in the book makes this a very suitable text to serve as the foundation of seminar discussion for students. It tears down old (often unnoticed) metaphysical prejudices, but is not able to erect the new building that, anticipated by many, still awaits the creative hand. Perhaps this is because his naïve theory of representations is insufficient to capture the whole of the knowledge process; in particular, it cannot exhaust the relation concept as Schlick believes. Perhaps it is also a result of the fact that Schlick's epistemology, which can be called psychologistic, is not psychological enough, i.e., it does not dive into those depths where the psychological is transformed into the logical. Maybe the part of psychology that cannot be reduced to neurophysiology, as Schlick would like it to be, can provide the extension that he cannot deliver.

[7] But as a thorough, organizing, and cleansing achievement, as a valiant anti-metaphysical act, as a signpost that names things properly, the book should receive its due attention.

2

Einstein's Theory of Space

Dynamics, or the theory of motion, is the science of the temporal passage of spatial events. At least this is how this study has come to be defined. But is it really true that we can arrive at a clear understanding of dynamics using this definition?

The naïve understanding is satisfied with this explanation. Indeed, what is so clear and simple as *space* and *time*? *Space* is what we see with our eyes, and *time* is what we feel as everything is passing by, one thing always after the other. But is this true? Who has ever seen space? I mean that one can only see *objects* in space and that they stand in the particular relations we call "in front of," "behind," "to the right of" and "to the left of." We coordinate every object with a place in space; but to speak of the *space* itself, we then have to mentally extract all of the objects. That is a very broad abstraction. How do I know that this *space* exists when all bodies are removed? Not through *experience*, since all observations

Translated from "*Die Einsteinsche Raumlehre*," *Die Umschau*, vol. 24, no. 25 (1920), pp. 402–5.

refer to those *real things* and their respective distances can only be defined with respect to the things around them. It is therefore a peculiar construction in which we embed things such that it is attached to them but can never be observed, and, unlike forces or heat that will make them glow, has no effect on them, yet it dictates far-reaching laws. We know all of these laws – two things cannot be at the same place at the same time; and for a body whose edges all form right angles, no more than three edges can meet at a point; and if I draw a straight line on this piece of paper and a point somewhere next to it, I can draw only one line on the paper through this point which does not intersect the first, etc. These are after all very strict limits that the things must obey; for example, why can't *four* edges meet at a corner? Or why will my right glove not fit on my left hand? After all, they have the same number of fingers and are the same size and even though they are exactly the same in that I can map them point by point onto one another, it still will not fit. And what determines these limitations is space as an invisible empty building with rigid partitions of which we have no more real concrete knowledge than of the dark side of the moon which has also never been seen by humans. And we call this artificial entity a clear representation? This is how we want to define an exact science?

And how about time? How quickly does time actually flow? Everyone has at some time observed that it depends entirely upon one's frame of mind; sometimes time flies by at a furious pace, and sometimes it creeps along and almost seems to have stopped. Consider a time at which every machine on Earth was suddenly quiet, all water stopped flowing, all people and animals, or generally everything that moves, would be completely still. I lie on the ground and stare up at the sky, no cloud moves, no breeze is blowing. Would time still be flowing? Yes, since I still breathe and feel the rhythm of my pulse. Consider these also to stop. Now I can still feel time in the sequence of my own thoughts, since their eternal change creates a permanent sequence. Now I also try my best to bring this process to a standstill – is there still time?

No, it makes absolutely no sense in this circumstance to speak of time. We only have temporal orderings when there are *real processes* that pass by, and we only have spatial orderings when we have *real things* to see and measure. And here we find our first result: space and time are not *things* and cannot be *perceived*, rather they are *empty schemata* through which real things are ordered and only when applied to real things do they achieve any determination.

If we want to be serious in our use of this knowledge, however, then we must radically revise our propositions about these schemata. We have always accepted all such propositions as self-evident. But now we must try to separate the propositions that only relate to the *schemata* itself from those that relate to the *real things*; but that is in no way simple because we have always been accustomed to treating propositions about spatio-temporal relations in the same way as any other scientific statements.

Let us begin with a simple proposition of daily life. We have two teacups in front of us that are of exactly the same shape. I place them rim to rim and conclude that they are exactly the same size. I now take one cup and move it to another table. The cups are naturally still the same size, right? – but wait a minute, we used the word "naturally," and that is always very suspect. We always say "naturally" when we come to those self-evident truths that under closer examination appear quite strange. Therefore, we maintain the opposite. The distant cup, we claim, is twice as big as the first. But this can be easily checked – we simply bring the other cup over and put them next to each other as before; here we verify again that they are the same size.[1] This verification is exactly right – but now if the second cup were moved to the table would it double in size? How would I decide *this*? Let us assume that a law exists whereby every body that is brought to the other table doubles in size and returns to its previous size when brought back; there is no possibility at all for us to verify this law. Many say that experience teaches us that objects remain the same size when moved to a different location. But we see that experience cannot teach us this because even if the contrary were true, we would not notice it.

This is another very important discovery. To illustrate, consider the following well-known statement of physics: the energy of a closed system of bodies can neither increase nor decrease. This proposition is perfectly sensible because it can be demonstrated experimentally; it would be very easy to verify whether, say, the energy of a closed system over time always gradually decreased (which is perfectly conceivable). And the only reason why the energy principle has any physical meaning is because *its contrary* is liable to *verification*. It is just for this reason that it is meaningless to say that two bodies at distant points in space are the same size, because we could not observe if they were unequal. We can only say that *if the bodies are placed next to one another they are of the same size, and when placed together at any other point in space they will still be the same size; but if one were here and the other were there, I can just as well say that one is twice as large as say that they are the same size.*

[1] We have no other way to verify that the sizes are the same. The objection is not that both cups appear the same size to the eye when separated. I can, of course, also make the teacup appear to be the same size as the house across the street by placing it in a suitable position. We only have an objective control if the cups are in direct contact.

This insight leads us to a very wide-ranging conclusion. The previous physical propositions are all posited under the assumption that it objectively makes sense to say that two bodies at different points in space are of the same size. Yet we found that this assumption is false, the assumption of equality could just as well be replaced with its negation, hence, we must demand that the contrary assumption leads to the same *physical laws* as before.

The *physical laws* themselves must not change under arbitrary choice of spatial proportions, or else we will be left once more in the position of having to privilege a particular system of measurements as objective: namely, the one that yields the "true" physical laws. *Einstein's great achievement is to have extended the consequences of this thought to all the foundational laws of physics.* The remark has occasionally been uttered in the past that the choice of basic ratios is arbitrary. But Einstein is the first to come to the conclusion that the *physics* must undergo a corresponding change, and further the first to have accomplished the necessary scientific revisions mathematically. Here lies the remarkable scientific contribution in whose time we are privileged to live. We will consider the physical consequences of the theory of relativity in an essay to be published in a subsequent edition of this journal.[2] First, however, we must clear up some conceptual difficulties.

Physics is a *quantitative* science; its great results lie in the fact that its theories can always be tested through *numerical measurements.* Do we surrender this advantage with our considerations? What is left then of the quantitative relationships if the geometric relations are considered as fluid, if the elementary basis of all measurement is eliminated?

[2] Chapter 6, "Einstein's Theory of Motion," in this collection.

It is clear that we must put new quantitative relations in place of the old statements, if we want to bring about scientific results at all. Another experiment with our teacup will show us the way.

The cups are to be placed in different corners of the room. Under these circumstances, I know that it makes no sense to say that they are of the same size. But consider, now, a different measurement. I measure the distance between the cups by repeatedly laying a meter stick down along the line connecting the cups until it reaches the far cup. We can lay the meter stick down, say, seven times in this way. From my earlier perspective I would say that the geometric distance between the cups is seven meters. I know that I can no longer say this. The distance could just as well be, say, ten times as large, because I cannot tell whether the meter stick (and my own body for that matter) expands when I move it from place to place. But wait, here we have come across an important fact, namely, that even if it does expand by a factor of ten when transported, the meter stick can be laid down seven times; at last we have an objective statement. *The factor 7 has a real meaning here*; it even holds if we were to think of space as arbitrarily deformed, because it would always deform the meter stick with it; here is finally an objective *numerical measurement.*

Note the distinction. We now refer to distance as a *numerical relation between material things*; it is a verifiable number coordinating the material teacups, on the one hand, to the material measuring rods, on the other. Formerly, distance was a *geometric* concept separated from the objects and justified by the nature of space. Now we think of distance as a *physical* concept that only contains an assertion about real things; distance is a *physical property* of objects, just like temperature or magnetism. And with this *physical* distance, we must now construct a new quantitative physics.

Here is an example of how the new notion of distance gives rise to objective statements. Consider a very large semicircle made out of wire about the size of a town. We take our measuring rod and measure the length of the arc of the semicircle by repeatedly placing the rod along it. We find that we can lay the meter stick down about 15,000 times. Then we take our rod and measure the distance along the diameter; we divide the number by two to get a radius of about 4,770. These are both *physical* distances in the sense of our new definition, as they signify only the numerical ratio of real bodies. Dividing the two numbers gives us the well-known number $\pi = 3.14$. This *empirical* measurement confirms the well-known geometric relationship between the circumference and the radius of our semicircle.

Next consider the experiment repeated under different conditions. Let a strong force field exist in the region around the center-point of the semicircle that contracts all bodies to the same extent. The force drops off quickly enough that its effects are negligible over the rest of the radius. The deformation of the semicircle will therefore be negligible. We now take the same measurements as before. We again measure the physical length of the circumference to be 15,000. For the physical length of the radius, however, we get a larger number, say, 5,300, since the measuring rod contracted when it passed through the central region. The ratio of the circumference to the radius is now no longer $\pi = 3.14$, but a smaller number. In this case, we therefore *no longer* maintain the usual circular relation between the circumference and the radius.

Let us recapture our line of thought. Relativistic considerations forced us to abandon the *geometric* concept of length and replace it with a *physical* one. In reviewing the second circle measurement from the old perspective, one would say: the number for the radius, 5,300, is not the "true" length, but is distorted by the effects of the force field at the center point; we must use it and the number to calculate backwards to arrive at the "true" length and the "correction factor for measuring rods." Now we know: there is no true length other than the number of times I have laid down the measuring rod; this *physical* distance was our last hope for a quantitative relation, and it is the only thing we can talk about. With our new notions of distance and length, we *no longer* have the circular constant π. *Our physical geometry* is no longer "Euclidean."

We observe that our new concept of length leads to a result with many consequences. In fact, empirical measurement is placed in the position of being the ultimate arbiter of whether Euclidean geometry holds in reality. Indeed, the semicircle experiment could be performed to see which result is obtained. Admittedly, the result would be very inaccurate, and science would thereby prefer another method of demonstration; *but the described thought experiment should show that we are in a situation where, in principle, the validity of geometry is a matter of empirical determination.* And we can now say, based upon very precise astronomical measurements, that Euclidean geometry no longer applies, that wherever the forces of gravitation occur we will obtain measurements like those in the second mentioned case. Gravitation has the effect of destroying the Euclidean structure of space.

We will explain why it is the force of gravitation, and no other force, that plays this strange role in the previously mentioned forthcoming essay. Here we only want to point out a strange result. We assumed the equal status of all measurements and fundamentally followed a *relativistic* way of thinking. Nevertheless, our considerations have led us to an absolutely *objective* determination of the structure of space. Relativity does not mean the abandoning of a judgment, but the liberation of the objective sense of knowledge from its distortion through our subjective nature.

3

Reply to H. Dingler's Critique of the Theory of Relativity

Hugo Dingler recently made the claim that the conceptual foundation of the theory of relativity is false.[1] It will be shown that Dingler's argument contains serious flaws and is thereby untenable.

1. Dingler refers to the well-known example of the accelerating railway car. Although by now this example has become rather trivial as a result of its frequent discussion and was completely clarified by relativity theorists,[2] particularly in connection with Lenard's repeatedly proposed misunderstanding, we will consider it again here for the sake of completeness. Dingler explains that the braking railway car would not be comparable to one in uniform motion because in the first case the (negative) acceleration does

Translated from "*Erwiderung auf H. Dinglers Kritik an der Relativitätstheorie,*" *Physikalische Zeitschrift*, vol. 21 (1921), pp. 379–84.

1 See volume 21 of this journal, p. 668, 1920.

2 See e.g., A. Einstein, "*Dialog über Einwände gegen die Relativitätstheorie,*" in *Naturwiss*, vol. 6, pp. 700–701, 1918; and A. Kopff, "*Das Rotationsproblem in der Relativitätstheorie,*" in *Naturwiss*, vol. 9, 9ff., 1921.

not directly act on the objects in the car, but rather will be first transferred to them by elastic forces. Only in a "real" gravitational field, e.g., one that is parallel to the rails and brought about by a single large mass, can we have complete equivalence; but here again there is no perceptible difference between uniform and accelerated motion if we discount frictional forces. The latter is true, but the standpoint of Newtonian theory does not provide an adequate explanation of it because the accelerated train in this example represents a system (in free fall), for which the Newtonian laws apply, although it itself experiences an acceleration relative to the collection of uniformly moving inertial systems of the cosmos. For Newtonian theory, it is by sheer coincidence that the force of gravitation will be exactly compensated everywhere in this system by an inertial force so that a "virtual" inertial system results; the same coincidence makes the inertial and gravitational masses exactly equal.

Now, Dingler will surely deny this coincidence because he has elsewhere given another "proof" for the equivalence of inertial and gravitational mass.[3] In reproducing his proof, I will show how Dingler profoundly misunderstands this problem. He considers a scale from which two equally heavy masses are suspended and balanced and concludes the following:

"A scale with both masses resting in equilibrium can be understood in the following way: the masses both receive a certain equal, downward acceleration from the Earth's gravitation. This acceleration is not observed because there is an upward acceleration of precisely the same amount and in precisely the opposite direction.

[3] See H. Dingler, *Grundlagen der Physik*. Berlin, 1919, p. 94.

Since they cancel each other out, the scale is thereby at rest. The upwards pull is not caused by gravitation, but by some local force (such as my hand, etc.). With respect to the tension, which gives both masses an upward acceleration and affects the pivot point of the lever, both masses are inertial and not gravitational masses. However, this scale, of course, cannot tip (otherwise we would not have equal masses as was presupposed), as a result the scale remains in equilibrium and it can be seen that both masses also have equal inertia."

This argument hinges upon the strange claim that the elastic pull exerted by that which is suspending the scale is transmitted through the scale's central beam to pull both masses upward compensating for gravitation in such a fashion that is proportional to the inertial mass. Dingler will have a hard time explaining how this is possible, since inertial mass is defined in terms of resistance to changes in motion and nothing at all is said about motion. (It seems that the cause of his confusion is an ambiguity in the uses of the word "acceleration," first in the kinematical sense as an increase in speed and second in the dynamic sense as the strength of the gravitational field.) In truth, the elastic pull is, of course, determined by the heavy mass, whose pull it has to compensate. Dingler's proof of the equality of inertial and gravitational mass, therefore, does not include the inertial mass at all.

But Dingler's criticism of the railroad problem obscures the real issue. If the train is decelerated by friction, the acting forces are not proportional to the mass. That these forces do not have an effect on the unsecured objects in the car is caused by the objects' inertial resistance; were the inertia arbitrarily small, an arbitrarily small force applied to the bodies would suffice to change their motions. The observable effect 'the horizontal[4] displacement of the bodies, the pressure on the side wall' is therefore caused by the inertia of the bodies and also quantitatively determined by it. If inertial effects should appear, the acceleration cannot be created by the mass-proportional forces. Therefore, it is completely senseless to want to eliminate frictional forces from the problem; they (or any other elastic forces) have the task of bringing two systems of bodies into relatively accelerating motion. With mass-proportional forces, this is not possible. (Lenard's objections also completely fail to address this problem.) Only in this way does one encounter the sort of situation that the theory of relativity considers because it wants to show that in two relatively accelerating systems of bodies the distinction between inertia and gravitation is arbitrary.

Further, the theory of relativity also posits the following oft-proven assertion which not even Dingler can dispute: Assume the (stopped) train stands still and a gravitational field[5] appears, the same effects come about as when the railway train is (negatively) accelerated in a gravitation-free system. The validity of this assertion is indubitable; there are no phenomena, even those outside of the train, which could not be explained in both manners. The theory of relativity further concludes: Let two different events E_1 and E_2 have the same effect W; then in observing an occurrence of W, I may not conclude that only E_1 could have occurred. But if one argues like Dingler (and Lenard as well) that the effects in

[4] Since the vertical component of downward falling bodies results from the Earth's gravitational field, and this effect is not of interest here.

[5] This gravitational field applies not only within the car, but also in the surroundings of the train. The question of the source of these gravitational fields need not concern us here; it is well known that they can be deduced from mass effects just like all other gravitational fields.

the car prove that the train is moving, then one is led to these fallacious conclusions.[6] It is the great achievement of Mach[7] to have pointed out this fallacious conclusion; he adds that in this case E_1 and E_2 must be seen as identical. Only the relative motion of the objects is real.

2. Dingler wants to determine the state of motion of the inertial system of the cosmos, apart from the rotation (which will be discussed below), by relating it to the totality of the fixed stars. The Galilean laws of inertia hold only in those systems that are obtained through gradual astronomical approximation. Here Dingler fails to notice the assumption that the calculations involving various fixed stars based upon the application of Newton's laws always lead to identical inertial systems. This assumption, which also occurs in Newtonian theory, cannot be proven *a priori*, and to the contrary seems unlikely from the start. It is the purpose of relativity theory to replace this requirement with a hypothesis that is much more intuitive and experimentally well-supported: for every point event there is an associated inertial system, but in general for every world point there is a privileged system. This is Einstein's Principle of Equivalence: the gravitational field can always be locally "transformed away." In place of the integral principle of Newtonian theory, we therefore put a differential principle; and while this seems to be experimentally well-grounded because of the equivalence of inertial and gravitational

[6] It cannot be claimed that E_1 is preferable because it is simpler. There are three possible explanations here: 1. E_1 occurred. 2. E_2 occurred. 3. It is arbitrary to consider E_1 or E_2 to have occurred. That the last assumption is the simplest, i.e., that it explains within the framework of a uniform theory as many facts as possible, has been proven through Einstein's requirement of general covariance.

[7] E. Mach, *Die Mechanik in ihrer Entwicklung*. Leipzig, 1904, 5th ed., p. 252. It can also be found in A. Kopff, loc. cit., p. 10.

mass, the integrability is restricted to particularly simple cases that can only rarely be realized and even then only approximately.

Dingler wants to use Foucault's pendulum to fix the orientation of the inertial system with respect to rotation. He considers this to be an entirely novel result that he developed over the last fifteen years (this journal, loc. cit., pp. 670–2). This is quite surprising as it is nothing more than Newton's idea that rotation implies absolute space. Whether the absolute orientation is defined in terms of the disappearance of the centrifugal force or in terms of the plane of the pendulum, the difference is not worth mentioning; the inertial forces always remain distinguished as absolute.[8] Dingler's view is therefore a hybrid of Newton and Mach; with respect to translation, the state of motion of the inertial system of the cosmos is determined by the masses of the fixed stars, with respect to rotation it is determined by absolute space. This solution to the problem of motion cannot be called rigorous, and neither can a theory of knowledge that would permit such a solution. The error in this solution is immediately exposed if general covariance is seen as empirically confirmed because then it means, just as above, that we can conclude from the observation of W that E_1 occurred.

3. Dingler wants to define simultaneity through the transport of clocks that have been locally synchronized. He grants that this definition of simultaneity is based upon certain suppositions, but

[8] Foucault's pendulum is, of course, very clearly explained from the standpoint of the theory of relativity: the pendulum, lined up with the North Pole of the Earth, would from the Earth's perspective also change its plane if the Earth is assumed to be at rest and the fixed stars rotate about it. The fixed stars "drag the plane of the pendulum with them."

he formulates the necessary axiom very unclearly and then asserts that it is also assumed by the theory of relativity. The condition he sets out is that "a mere movement of a clock does not alter it" (ibid., pp. 671–2), and then charges that the theory of relativity wrongly assumes that clocks move uniformly, although every clock at rest could be considered to be in motion. He ignores the fact that all speculation concerning the change in rate of a single clock is vacuous and that only comparisons with other clocks leads to observable phenomena. The axiom must therefore read: "When two neighboring clocks are synchronized, they will maintain their synchronization if they are transported to a different place along separate routes." This is an empirically verifiable statement. The theory of relativity considers this to be false, asserting that, upon their second meeting, the clocks will in general not remain synchronized. Admittedly there is no direct proof of this, but very many indirect demonstrations; yet Dingler gives no other support for his axiom than to say that before Einstein everyone believed it. But even the opponents cannot say that the theory of relativity implicitly assumes the very axiom that it explicitly refutes.

4. Dingler does not recognize that the theory of relativity has been empirically confirmed, since he has stated that it remains to be determined "whether the theory of relativity is capable of empirical confirmation or refutation, or indeed if any theory can be" (ibid., p. 673). This is one of the many cases in which Dingler, through a faulty formulation, distorts a perfectly good idea. If I have demonstrated that a body is square, then it has been shown that square bodies exist and this body is one of them. It is needless to demonstrate the general theorem beforehand, since this demonstration is contained in the proof of the single case.

Rather, the problem is whether the proof of this individual case is acceptable; therefore, it is necessary that the criteria for a valid proof are set up and it is determined whether the proof in question satisfies these criteria. Dingler is right in saying that these criteria depend upon logical requirements and in this way logical requirements are present in physics.[9] It is, however, a grave mistake to claim that the theory of relativity ignores these requirements. Quite to the contrary. The criteria I propose for the validity of a theory are:

1. The theory must be free of contradictions.
2. The theory must the simplest of all of those that concern the same phenomena.
3. The interpretation of the empirical data must follow from the normal process of physical induction.

All three criteria are satisfied by the theory of relativity. Concerning the second criterion, no final judgment is possible; but as of yet no simpler equivalent theory has been posited in spite of many attempts. For the significance of the third criterion, see my above-mentioned book (pp. 60–64).

5. Dingler advocates the view that no empirical decision can be made about the validity of geometry. This is correct in one respect: if one wants to hold onto Euclidean geometry, then one can alter the physics so that this demand is satisfied. One must, however, give up the epistemologically fundamental demand that natural rigid bodies satisfy the congruence axioms of the theory.

[9] However, it does not follow that these requirements are completely arbitrary as Dingler maintains in his above-mentioned book. For a discussion of these questions, see my book: *Relativitätstheorie und Erkenntnis a priori*, Springer, 1920. Especially sections V and VI.

Dingler disputes this alternative and instead gives the following proof: in order to prove that Euclidean geometry does not apply to natural rigid bodies, I must assume Euclidean geometry in my empirical measurements; in this way the argument is circular. The objection is therefore raised according to the previously mentioned first criterion.[10] I have already pointed out the fallacies in this argument in my discussion of the Kantian philosophy, and Dingler's objection is no different than the conclusion of the Kantian philosophy of the *a priori.* I now repeat the content of my thoughts as formulated by K. Grelling in a recent conversation. Suppose: "A is B." I prove through logically correct operations and lemmata that A is not B. It thus follows that in fact A is not B because the contrary assumption leads to a contradiction. We only have to require that the assumption "A is not B" does not lead to the consequence that "A is B" or else a real paradox results.

Now this is in fact the method of argumentation employed in the theory of relativity. We will illustrate this with an example that indeed does not prove the falsity of three-dimensional Euclidean geometry, but shows something similar for the four-dimensional space-time manifold. The equivalence of inertial and gravitational mass is proven under the assumption of Euclidean geometry (in three- and four-dimensional space); as in measurements of the torsion balance, this geometry underlies the design of all measuring instruments. Another confirmation of the equivalence of the masses lies in the fact that Newtonian gravitational theory at least approximates real space, so this law also works with the

[10] 1. Theories must be free from contradictions, 2. Theories must be the simplest of all of the options that account for the same phenomena, 3. The interpretation of empirical data must follow the normal induction rule for physics.

equivalence of the masses; admittedly this confirmation is not as precise as that of Eötvös. From the equivalence of these masses it now follows that for all space-time points there exists a local system in which no gravitational field is detectable, namely the rest system of a freely falling point mass. For space-time points with different Newtonian gravitational field strengths (i.e., inhomogeneous fields), these local inertial systems are in accelerated motion relative to each other. This is now joined by a second supposition that has been experimentally proven again given the assertion of Euclidean measurement proportions: that special relativity holds in such gravitation-free systems. We must add that this also holds in the case of coordinate systems of arbitrarily small sizes so that the absolute size of the spatial dimensions does not influence the validity of the special theory of relativity (homogeneity of space). In the local inertial system, the Minkowski line element is therefore given as $ds^2 = \Sigma dx_\nu^2$. It once again follows that this expression fails to hold for all systems in accelerated motion with respect to it, where the coordinates are defined through rigid rods and clocks at rest (we call these "natural coordinates") because every non-linear orthogonal transformation changes the line element to $ds^2 = \Sigma g_{\mu\nu} dx_\mu dx_\nu$ If we now try to cover the entire space with a coordinization in which each point satisfies the Euclidean measurement rule $ds^2 = \Sigma dx_\nu^2$ then the individual parts of the spatial coordinates must be in accelerated motion relative to each other because at every point they must correspond to the local inertial system. It thereby follows that the coordination in general cannot be realized by the method of rigid rods. By that it is proven 'because the converse also holds since only the local inertial systems follow Euclidean measuring rules' that a coordinate system realized through rigid rods and clocks at rest does not obey at all points of space four-dimensional Euclidean measurement rules. We carry out therefore the following deductive schema:

Supposition a: The validity of Euclidean geometry in natural coordinates.

Leema b: Equivalence of inertial and gravitational masses.

Leema c: Validity of the special theory of relativity in such coordinate systems in which no gravitational potential is observed; also for arbitrarily small regions.

Conclusion d: Invalidity of Euclidean geometry in natural coordinates.

This contradiction illustrates the logical procedure described above. Of course, the reverse schema does not hold, i.e., from the supposition of non-Euclidean geometry, the validity of the Euclidean does not follow, and so we do not have a paradox.

This in no way proves the invalidity of Euclidean geometry for natural coordinates; rather, this invalidity necessarily follows only when lemmas b and c are assumed. We are only trying to show that the proof is at least logically possible, because lemmata b and c are at least logically admissible assumptions.

We are thus given two possible solutions to the contradiction: drop either a or c. Dingler would probably prefer the latter; but he would contradict an empirically well-founded and logically permissible physical result, the special theory of relativity. (Even Dingler has not been able to raise a tenable objection to the special theory of relativity.) What I want to establish here, however, is that Dingler cannot justify his solution by showing that an alternative solution is impossible. The theory of relativity also gives a permissible solution to the contradiction as we will now demonstrate.

The theory of relativity considers b and c to be so empirically sound that the invalidity of Euclidean geometry is considered proven. A physical result is thereby produced, although negative in character. However, this is insufficient for the theory of relativity which seeks to positively validate a certain geometry, namely Riemannian geometry. And here we encounter the larger logical problem which is not covered by Grelling's argument schema: if a is dropped, can I still make use of b and c in the demonstration of Riemannian geometry given that they entail a? The problem can be put in this way: after having shown that "A is not B," I want to prove that "A is C." However, I only have theorems at my disposal (namely b and c) that depend upon the assumption "A is B," hence assume that "A is not C." Can I argue as follows: supposing that "A is not C," I prove that "A is C," but does this mean that "A is necessarily C"?

That would be a fallacy. But this is not the method of argumentation employed by the theory of relativity. Upon closer examination of theorems b and c, we find that they do not necessarily assume a, as they already hold if we limit consideration to a small region of space, e.g., in the dimensions of the torsion balance that confirms the equivalence of gravitational and inertial mass, or the electron tube in which the electrons move according to the laws of special relativity. The exactness of the measurements in our empirical results does not allow a distinction between Euclidean and non-Euclidean geometry in small dimensions. Indeed the deviation of the equivalence of masses from the validity the Newtonian laws rests upon the validity of Euclidean geometry in astronomical dimensions; but it is exactly this derivation that is faulty, and it remains an open question whether one wants to explain this approximate equality by the merely approximate validity of Euclidean geometry.[11] The theorems b and c are therefore compatible with the validity of a Riemannian geometry of small curvature. This gives us the following argument procedure:

Supposition: *A* is *B* or *C*.
Auxiliary theorems based upon this supposition.
Conclusion: *A* is *C* and not *B*.

This is a permissible demonstration for the proposition "A is C."[12] I have elsewhere referred to this method as the "process of successive correction";[13] it is very characteristic of modern physics and allows an unforeseen expansion of the principles of scientific discovery. It is also, for example, applicable to quantum theory, where the discrete structure of energy is verified with the aid of experiments that assume energy to be macroscopically continuous because every computational analysis of the experimental set up naturally works with the concept of energy in the old sense, as a continuous magnitude. In this case, the conclusion

[11] The theory of relativity asserts therefore that an accurate confirmation of the equivalence of masses from astronomical data using Euclidean geometry and Newtonian laws can never be produced; but this is trivial, since it merely says that even the Newtonian laws do not hold. A quantitatively exact confirmation of the principle of equivalence in astronomical dimensions would require the use of Einstein's equations for the gravitational field. Such a confirmation would in no way be vacuous, since it is well in line with the general process of physics to demonstrate the validity of an hypothesis by first assuming it and then confirming its unquestionable feasibility through repeated experiences. The astronomical confirmations of the theory of relativity thereby constitute proof for the entire theory as much as for the auxiliary theorems b and c.

[12] Additionally, the compatibility of the theorems b and c with C poses no problem for the Grelling argument scheme because nothing here is changed if the auxiliary theorems depend upon "*A* is *B* or *C*."

[13] Loc. cit., pp. 66–68.

follows in completely the opposite direction, from larger to smaller dimensions.

Logico-epistemological objections to the theory of relativity are therefore unsuccessful.

6. Like Dingler, I have long been of the opinion that it is necessary to give an axiomatization of physics, indeed the current state of this science urgently demands such an axiomatization. It is not possible, however, to dismiss the already existing results from the pre-established physics that have found a magnificent extension in the theory of relativity as a part of "the old physics proper." I consider it extremely regrettable if the axiomatic process is to be discredited from the start by misinterpreting physical considerations and fallacious reasoning.

4

A Report on an Axiomatization of Einstein's Theory of Space-Time

A clarification of the fundamental concepts of the theory of relativity will only be possible if the tenets of the theory are presented in axiomatic form. The axioms contain the fundamental facts whose existence justifies the theory; in principle, they are *empirical* assertions capable of experimental verification. In addition to these are the *definitions* through which the theory's conceptual content is constructed. These, in contrast to the axioms, are *arbitrary* forms of thought, capable of neither empirical confirmation nor refutation. Their arbitrariness is only constrained by certain well-understood *logical* demands that are required for a *scientifically appropriate* system. They must be *univocal*; but moreover they must lead to a scientific system characterized by certain properties of simplicity. Whether they fulfill these demands is not solely a matter of form, but depends upon the validity of the *axioms*; the preferred properties of the conceptual system derive from the facts

Translated from "*Bericht über eine Axiomatik der Einsteinschen Raum-Zeit-Lehre*," *Physikalische Zeitschrift*, vol. 22 (1921), pp. 683–7.

laid down in the axioms. The results of the systematic derivation of all the theorems of the theory from the definitions and axioms, viz., "ordine geometrico," are twofold: on the one hand the discovery of those facts without which the theory could not exist and whose validity completely proves the theory, and the other hand the discovery of its logical structure which gives the theory its legitimate place as a scientific theory.

The axioms fall into two classes: *light axioms* (I–V) and *matter axioms* (VI–X). The first are the assertions concerning the physical properties of light independent of any relation to material objects. It will be shown that only through these axioms is it possible to construct a *complete theory of space and time.* The matter axioms imply the identity of the developed "light geometry" with the space-time theory of rigid rods and clocks. The most important result of this investigation is that not only is this separation possible, but further that even without the validity of the matter axioms, whose empirical confirmation cannot be completely carried out, the theory of relativity is a *valid* and complete physical theory.

We first develop the axioms for the special theory of relativity. We conclude our report with the extension to the general theory.[1]

1. *Axioms of time order.* We first define the time order at a point. A light signal sent from a point *A* to an arbitrary point *B* (which may be moving) is reflected and returns back to *A*.

Definition 1. The departure of the signal from *A* is called "earlier" (written $<$) than its return to *A*.

Temporal neighborhood is thus defined for each individual point. In this way, the determination of "simultaneity at a single point" is assumed to be possible; this is a philosophical problem

[1] A detailed publication of the entire investigation is forthcoming.

and requires an investigation that will not be undertaken here. The definition of succession will be supplemented with five axioms, which make it univocal and define a continuous infinite temporal sequence for the point. The most important among these axioms read:

Axiom Ia. If the event $E_1 < E_2$, then there can be no signal chain such that $E_1 > E_2$.

Axiom Ib. For any two events E_1 and E_2 at A there is always a signal chain such that either $E_1 < E_2$ or $E_1 > E_2$.

Axiom Ic. For all events E, there is an event E_1 and an event E_2 such that $E_1 < E < E_2$.

Transferring the time sequence of a single point to another is established via light signals. Since the definition of the time order at the other points proceeds according to the aforementioned process, an additional axiom is required.

Axiom II. If two light signals are sent from A to B, the time order of their departure from A is the same as the time order of their arrival at B.

These axioms are also valid for points in relative motion.

2. *Axioms of the comparison of time.* The notion of a "system time" must now be defined. Which points are relatively at rest must have been previously defined for this. We introduce the auxiliary concept "clock," although nothing at all is asserted about material clocks; in particular, the concept is valid for any arbitrary temporal metric.

Definition 2. A clock is any mechanism that coordinates each event to a point according to the sequence of real numbers.

Definition 3. Choose at A an arbitrary clock interval as the unit of measure of time. Find for each point P, all of the points (in arbitrary states of motion) for which the signal time (measured with respect to the selected unit) has the constant value APA. This point system is called a "reference system related to A."

Axiom III. It is possible to choose a time interval in A such that the resulting point system related to A is a reference system related to each of its points.

Definition 4. A system satisfying axiom III is called a "normal system"; a clock thereby calibrated to the system is called a "normal clock."

The so-defined point system is not yet univocally determined; a linear expansion is still possible where each point moves out radially from A with a constant velocity proportional to the distance from A at time $t = 0$. The temporal metric is thereby not determined up to a linear function as uniformity requires, rather it remains possible to choose a logarithmic function for time. But this will be impossible when axiom V is included; we will carry this out in definition 7. Provisionally, however, we do not make use of this further limitation; in the following, clocks will be understood to be the normal clocks of definition 4.

Within a normal system, the Einsteinian definition is used to synchronize all clocks:

Definition 5. A signal is sent from A at time t_1 reflected at B and returns to A at time t_3; the signal is received at B at the time:

$$t_2 = \frac{t_1 + t_3}{2}.$$

In a normal system synchronization is symmetric; i.e., when *A* is synchronized with *B*, *B* is synchronized with *A*.

Axiom IV. If from a point *A* of a normal system two light signals are sent simultaneously along a closed triangular path *ABCA* in opposite directions, then they will return simultaneously.[2]

Only now can we deduce that all clocks that are adjusted according to definition 5 are also synchronized with respect to each other, that the defined synchronization is not merely symmetric but also transitive. (I call relations with both properties "substitutive.") Herein lies the most preferable aspect of Einstein's theory of time. It can be derived solely from the basic empirical facts in axioms II, III, IV.

3. *Metrical axiom.* We define straight lines via light rays and congruence through:

Definition 6. If two signals depart simultaneously from *A* along the paths *ABA* and *ACA* and return simultaneously to *A*, then we call *AB* = *AC*.

Axiom V. It is possible to select a normal system with respect to *A* such that the metric is Euclidean.

Definition 7. A system satisfying axioms III, IV, V is called an inertial system; the points of such a system are said to be "at rest relative to one another," and the accompanying normal time is called uniform.

2 Einstein has already pointed out the significance of these axioms for synchronization in his lectures.

It can now be demonstrated that this definition is univocal. Thereby, points at rest with respect to one another are defined without the use of rigid rods; it is a "light rigidity" that is defined. Likewise, uniform time is characterized through the motion of light without the use of material mechanisms. This occurs likewise with the Einsteinian light clock, albeit that light clock is still based upon rigid bodies.

It follows from definitions 5 and 6 and from axioms II–IV that the speed of light is a constant of the system. The principle of the constancy of the speed of light is thus not merely another light geometrical definition, but also one that rests upon certain empirical facts. In order for it to hold for the metric of rigid measuring rods also, the matter axiom VII must be added.

4. *The Lorentz transformation.* The transformation connects uniformly moving systems which are defined by:

Definition 8. A point system K' moves uniformly with respect to a system K if all points of K' are displaced by the same amount along similar paths in equal time as measured in K.

It must also be assumed that axioms I hold in K'. It can then be shown that axioms II–V obtain. K' is also therefore a system whose points are at rest with respect to one another and it can therefore be shown that K in turn moves uniformly with respect to K'. The transfer of units of length and time obeys the following definitions:

Definition 9. The length of the unit at rest in K', measured in K, must be equal to the length of the unit of length at rest in K, measured from K'.

Definition 10. The unit of time in K' is to be chosen such that the velocity of light related to the unit of length projected according to definition 9 receives the same numerical value in K' as in K.

The Lorentz transformation can be derived from this. From the light geometric point of view therefore it requires no additional axioms at all. Its relationship to rigid rods and clocks is contained in the matter axioms VI–X.

5. *The matter axioms.* "Rigid bodies" and "natural clocks" can be completely defined independently of the metric. These are given through the material behavior of objects when the influence of the environment approaches zero in the limit.[3] The so-defined space-time metric becomes connected with light geometry via the following axioms:

Axiom VI. Points at rest with respect to one another (according to definition 7) can be connected by rigid rods.

Axiom VII. Intervals that are congruent by definition 6 in an inertial system are also congruent when measured by rigid rods.

Axiom VIII. The unit determined by rigid rods when transported into a uniformly moving inertial system K' is equal to that determined by definition 9.

Axiom IX. A natural clock runs uniformly.

[3] A detailed justification of this is contained in a forthcoming publication. [See Chapter 7 in this collection.]

Axiom X. The unit determined by natural clocks when transported into a uniformly moving inertial system K' is equal to that determined by definition 10.

These axioms completely fix the identity of the "natural metric" with the light geometry. With respect to their empirical confirmation, it should be noted that:

Axioms VI and IX will scarcely be disputed. That there exists a system for which VII holds will likewise not be disputed; but that this axiom holds in all systems in uniform motion is an immediate result of the Michelson experiment. It is not a question of possible interpretation of these truths, but merely a reproduction of observed facts; the interpretation 'the constancy of the speed of light in all systems' is fixed by the definitions in conjunction with the other axioms. Axiom VIII is not yet experimentally guaranteed; it only follows from the Michelson experiment that the transported measuring rods are of the same length in K' by definition 6. If VIII were not valid, then there would be a privileged system in spite of the validity of VII in all systems; this appears unlikely. Axiom X contains the second statement of the principles of the constancy of the speed of light, that is to say, that c is a constant of the system which has the same numerical value in all systems; this axiom is not yet sufficiently experimentally confirmed. Investigations into the transverse Doppler effect using canal rays will be important for this reason.

The pursuit of the complete axiom set necessary for the theory of general relativity cannot be carried out in the framework of this short essay. It is essential that the analysis of measurement in the special theory of relativity determines that of the general theory of relativity. Of these results, we state the following:

(a) Axioms I–IV, VI, and IX hold for static gravitational fields. These yield a uniform time and synchronization, which, according to definition 5, is substitutive. This time does not differ much at all from that of the inertial systems, lacking only the transformability of length and duration through rigid rods and clocks.

(b) In stationary gravitational fields (e.g., a rotating disk), I–III, VI, and IX hold. We again have uniform time, but synchronization is no longer substitutive, it is only symmetric. With this, the most important property of the synchronization is lost.

(c) For arbitrary gravitational fields we still have axioms I and II, if they are restricted to points that can be represented as point-masses. Herein lies the nature of the speed of light as a limit for arbitrary gravitational fields as well; light remains the fastest messenger. Also in arbitrary gravitational fields, we can interpret time sequences to be the same intuitively clear process as above, but without the metric and substitutive synchronization, which they have always been understood to possess. The process represents a specific choice of coordinates, but this can always be achieved. However, it is *not mandatory* to use this simplified interpretation.

Discussion

Lohr: The lecturer has used equal time intervals to determine points at relative rest. What sense should one make of *equal* time intervals at a point since a clock (a periodic process) is not given?

Reichenbach: First of all, we assume the existence of a clock with a completely arbitrary metric; it gives a provisional definition of the equality of the time units. Through the demand that the resulting point system be a reference system *for all of its points* (axiom III), the metric will be considerably restricted; this demand is not realizable for an arbitrary metric. With the addition of axiom V, the restriction is so stringent that the metric is fixed up to an arbitrary linear function, i.e., that time is uniform.

Richter: If the lecturer accepts his considerations as valid for both the ideal mathematical and the real world, he must show where a mathematical point (*A* or *B*) exists in the real world, and how he thinks of a mathematical point in reality. The mathematical point is not real.

Reichenbach: In reality there are only approximations. That is always so and not only for the present address.

Richter: Further, he must show how in reality a mathematically precise center position between two points *A* and *B* can be assumed for the purpose of observation. The ability to think of such things does not coincide with its possible realization. Here as well the entire question of Einstein's theory reduces to the question of the epistemological relation between subject and object. Otherwise it is not possible to assume that a center position between *A* and *B* exists. How can he do this in reality?

Reichenbach: I have exactly defined what lengths are equal, and it has nothing to do with subject and object. It is true that the possibility of approximate knowledge remains a philosophical problem. You may call this an axiom; but such an axiom stands above all of physics and so I have not mentioned it.

Richter: So in reality, we can only speak of reality as an ideal world. (Compare this to my work: *Entwicklung der Begriff von Kraft, Stoff, Raum, Zeit durch die Philosophie mit Lösung des Einsteinschen Problems*. Leipzig, Otto Hilman, 1921.)

Frank: To the lecturer's axioms must still be added the relativity principle to derive the theory of relativity. However, if this is assumed, then the axioms presented by the lecturer follow partly from it.

Reichenbach: As a matter of fact, the mentioned axioms only imply the rules of measurement in the theory of relativity, but not the covariance of all laws of nature; but this is not the problem at hand. The Einsteinian explanation can still in a certain sense be called axiomatic in that Einstein begins with the two demands: the relativity principle and the constancy of the speed of light. But these are very general formulations. It is precisely the point of analysis to determine what individual assertions they decompose into. The given investigation shows that only one part of the individual assertions requires relativity as an axiom (namely axioms VIII and X), that a second part only defines it (definitions 9 and 10), and that a third part does not deal with it at all. Only in this analytical way do we overcome blurred general assertions to reach a minimum of requirements, and only in this way may a foundation for an exact criticism be created.

5

Reply to Th. Wulf's Objections to the General Theory of Relativity

The objections to the general theory of relativity raised by Mr. Wulf in vol. 5083–84 of this journal are entirely based upon mistakes that have strangely crept into his discussion of the theory of relativity; a correction therefore seems justified. I will follow Wulf's train of thought; he considers whether the Ptolemaic account, with the Earth at rest, leads to contradictions with observed facts from *Einstein's* point of view.

(1) This view allows for speeds whose numerical value is greater than the speed of light. This, however, does not contradict the theory of relativity, since the number 3.10^{10} cm $\sec^{-1}$ represents an upper limit only for inertial systems. Superluminal speeds, in the strict sense of the word, are non-existent, since no body can be moved quicker than a light signal at the same location in space and time in an inertial system. Further, in a gravitational field, light is the fastest messenger. The numerical value of this

Translated from "Erwiderung auf Herrn *Th. Wulfs* Einwände gegen die allgemeine Relativitätstheorie," *Astronomische Nachrichten*, vol. 213 (1921), pp. 307–10.

speed is dependent upon the definition of time in a gravitational field. Hence, in a gravitational field one cannot simply apply the Lorentz contraction of the special theory of relativity. In no way does it follow from *Einstein's* formulae that the planets would be shortened in the direction of their motion relative to a coordinate system attached to the Earth.

(2) That the fixed stars, relative to a stationary Earth, all have the same 24-hour period of revolution is also no coincidence on this view. One cannot – and this is one of the most common errors – simply consider the coordinate system of the Earth to be equivalent to a coordinate system that is at rest relative to the fixed stars (statistically averaged). But it is *Einstein's* assertion that we must attach a corresponding gravitational field to every coordinate system and that only then does this equivalence arise. The following are therefore equivalent: on the one hand, the coordinate system of the fixed stars with a gravitational field whose density increases in a knot-like fashion only in the vicinity of the stars and then consistently vanishes elsewhere, and on the other hand, the coordinate system of the Earth with a gravitational tensor field which is continuous throughout all of space and which reaches extremely large values as the distance from the center of the Earth increases. The movement of the fixed stars can be explained from the second point of view in the following fashion: it is the lateral components of the tensor field, which are observable from the Earth as the Coriolis force, that cause the fixed stars to move in a circle; their field strength grows proportionally with increasing distance from the Earth and therefore the fixed stars obtain a large rotational velocity that increases with the distance. Indeed, it is for exactly this same reason that we see all bodies fall with effectively the same speed, the equivalence of inertial and gravitational masses;

the fixed stars fall freely in a gravitational field whose strength increases as one moves away from the Earth. *Thirring*[1] has shown that such forces do follow from Einstein's field equations.

(3) This also clears up *Wulf's* objections to the equivalence of the coordinate system fixed to a carousel. He believes that the rays of light coming from the sun do not participate in the rotation of the Earth relative to this coordinate system, that the carousel will be at rest with respect to the pencil of light rays from the sun, and that the Earth must turn itself away from it. However, this is mistaken because the gravitational tensor field indeed affects the light rays, since they have weight and are thereby rotated along with the Earth. What turns the Earth and the stars is exactly this gravitational field; one cannot think that this is the work of the carousel horse. The force of the horse is used solely by the carousel to protect its motion against the pull of the larger rotation; the horse thereby only has an effect on the carousel, which is the exact opposite of his opinion. Consider a fly on the wall of a container that is initially falling. The container falls under the effect of gravitation, and because of its own gravitational mass the fly would be expected to participate in this motion. However, if it climbs up the wall of the container with the appropriate speed, then it would compensate for the force of the gravitational field with its own strength; the effort required to move in this way depends only on its own mass and is entirely independent of the mass of the container. In the same way, there will be a gravitational field if the carousel is standing still and one considers a fictitious rotating coordinate system; but the carousel is also subject to the pull because the force of the horse is not sufficient. And even if the

[1] *Physikalische Zeitschrift* **19**, 33, 1918.

carousel is rotating uniformly and the horse is doing no work other than compensating for friction, the gravitational field is present and continually drags the stars around; the field is independent of the work of the horse.

It may still be argued: the gravitational field in the falling container example is generated by the Earth, but where does the gravitational field for the rotating coordinate system come from? It comes directly from the masses just like every other gravitational field,[2] only here the field is generated solely by the rotating masses themselves. This is no longer paradoxical as soon as we assume dynamic gravitational effects. This is no different from a moving electrical charge giving rise to a field which then affects the charge itself. The analogy is entirely apt. Consider an electron at rest in an inertial system; from another inertial system the electron appears to be in uniform motion. In the latter system, there must be a force that repels the charge; this force will be supplied by the changing field itself so that an external force is no longer necessary.

(4) In this way we can also deal with *Wulf's* final objection that it would have to take years for the carousel's force to reach the fixed stars and drag them along. It is not this force that affects the stars, but the gravitational field. And this arises entirely from the masses themselves through their mutual effects and takes no more time to be transmitted than any other dynamic effect.

(5) It appears futile from the start to try to refute the theory of relativity through such considerations. Indeed, it was exactly these conceptual problems that gave rise to it. What could be called

[2] *A. Kopff* has shown that Thirring's force can be traced back to the masses in this way. See also the detailed treatment of the problem of rotation by *Kopff* in *Naturwiss* **9**, 9, 1921.

into question is whether these conceptual demands are properly couched in the mathematical formulation of the theory. But it appears that after *Einstein* deduced the well-known laws of celestial mechanics to a higher degree of precision and *Thirring* and *Kopff* were able to solve the problem of rotation, little seems in doubt.

6

Einstein's Theory of Motion

It is strange that we wait to subject a science's fundamental concepts to a pointed critique until it has reached a certain degree of completeness. So it is that the concepts of space and time have only now found a final clarification through Einstein's theory of relativity after physics was already a mostly complete science. But we can now say that it deserves the glory of a real, exact, natural science since the hitherto unnoticeable veil has been lifted from its elementary concepts.

The concept of motion as a *change in position over time* combines the most important categories in physics; for that reason its investigation is the starting point for the solution to space-time problems.

The most simple conception of motion is the "*kinematic*." For example, if a billiard ball moves on a horizontal surface, its position is characterized by its distance from two adjacent edges

Translated from "*Die Einsteinsche Bewegungslehre,*" *Die Umschau*, vol. 27 (1921), pp. 501–5.

of the billiard table; that the ball is in motion is a function of these changing distances. We speak of the *kinematic* concept of motion because it consists only of changes in *spatial lengths*. But what would one observe if the billiard ball was portrayed as a fixed point and the billiard table underneath was moved? Again, there are changes in both distances. Assume that I see the billiard table through a cardboard tube, so that I can only survey the green surface. I observe that the distances of the ball from both edges change; can I now claim that the ball has moved? Certainly, but I can claim with equal right that the billiard table has moved and the ball has maintained its position, for if this is the case then I would have observed the very same change of position. Now the obvious objection is that one could put aside the paper tube and easily see whether it is the ball or the billiard table that is changing its position relative to the walls of the room. Absolutely. But does it not follow from this that perhaps it is the *walls of the room* that are moving? All such considerations only embed the problem in a larger area of space. One can see: *so long as the change in relative position is the only criterion of motion, all motion is only relative*, i.e., there is no absolute decision about motion or rest. The kinematic concept of motion inherently contains the principle of relativity. From this standpoint, Copernicus is no more justified than Ptolemy; it is arbitrary whether the Earth or the fixed stars are to remain at rest.

This insight is quite old. It can be found in antiquity; in modern times it was Leibniz who set it out in its most precise form. But it is known that there are *other criteria of motion* besides change of relative position; the kinematic concept of motion is not the only one. It was Newton who gave these criteria their first precise formulation. Consider a horizontally rotating disk. The motion can be detected kinematically from the changes of the distances between some marked point on the disk and the surroundings considered to be at rest. We again have kinematic "relativity." Consider, for example, a little human being living on the disk; he observes the continuously changing distances of points in the surroundings and will initially maintain that the *surroundings* are in motion while he is at rest. We are like such a being with respect to the rotation of the Earth. Nevertheless, there is another possible way to decide which part of the world is moving: through the appearance of *centrifugal forces.* The "disk person" would realize that it is harder to go from the edge to the center point of the disk than it is to go in the opposite direction. The appearance of forces is thereby also a characteristic of motion; through the effects of forces, motion changes from a kinematic to a mechanical problem. And therefore we speak of a new concept of motion. Not only the changing spatial distances, i.e., the kinematic problem, but also the *appearance of forces* belongs to this concept of motion; it is very strange that motion, as we believe it to be *presented by material things,* always seems to be connected to the appearance of forces. This gives rise to the interesting question of whether there is a relativity principle for the mechanical concept of motion as well.

Newton answered this question in the negative because he observed that the centrifugal force does not appear when the disk is at rest and the surroundings rotate. For example, he hung a water bucket from a rope and rotated it by twisting the rope and then allowed it to untwist. He thereby observed the following: at the beginning only the bucket rotates and does not take the water with it. Then the water gradually comes to rotate and displays now that well-known whirlpool appearance rising up the sides of the vessel with the center hollowed out. This phenomenon is the well-known result of centrifugal force pushing the water to the outside.

Then Newton grabbed the bucket so that for some time only the water rotated. He observed that it maintained its hollowed-out shape. Thereupon, he concluded in the following manner: when the bucket is rotating around the water we observe no effect on the water; when the water is rotating but the bucket remains still then we have an altered surface. Rotation is therefore not relative; given a rotation in "absolute space," one can determine its magnitude from the centrifugal force.

That this view is logically untenable has been shown by E. Mach. What sense does it make to speak, e.g., of the rotation of a disk if it is perfectly alone in empty space? Nevertheless according to Newton, one should be able to define its state of motion "relative to space." In this way, Mach has shown that the bucket experiment proves nothing because the sides of the vessel are far too small. The water in the bucket at rest moves not only in relation to the bucket, but also relative to the world; perfect inversion could only be achieved by having gigantic masses rotating around the water. These masses could produce a *pulling effect* on the water equivalent to the centrifugal force. Similar to the way in which a *moving* electrical charge induces a different force than a *resting* charge, *moving masses* can also generate a separate *dynamical gravitation.* Mach considered it to be possible that his view could ultimately be experimentally confirmed.

Mach, however, *never succeeded* in formulating a physical theory based on his view. It was Einstein who *first attained* it; in his general theory of relativity he has given *completely relativistic* solutions to the problems of motion. According to Einstein, it is *mechanically* equivalent for any system to be considered at rest, hence even the Ptolemaic worldview is *mechanically* tenable. In the Copernican system, the force relations are simpler and lead us to the Newtonian law, albeit only as an approximation; however, the complex trajectories of planets in the Ptolemaic system are the result of forces that no longer resemble those of Newton's law.

Therefore a generally relativistic theory already follows from a strict analysis of the old mechanics, and it is surprising that it was not developed long before Einstein in light of Mach's published remarks of 1882. But it must also be described as lucky that it was not previously attempted, since this sort of mechanics could not have led to the correct results because there is another significant idea that would not have been known. Considerations of time are missing from our notion of relativity; this correction was not possible without a third concept of motion that we now introduce.

Apart from mechanics, the concept of motion also plays an important role in another field: *the theory of electricity*. It concerns processes that are only intelligible if electricity is understood as moving; only the comparatively minor problems of static electricity (which for the layman is usually exhausted by picturing Leyden jars and elderberry seeds) can avoid the concept of motion. But all technical applications derive from the *movement* of electricity that we know as electric current. In the last couple of decades we have considered a second phenomenon: the movement of the *electric wave* that transmits electrical energy through space and finds a surprising application in wireless telegraphy. One would not think that the concept of motion in the theory of electricity would be the source of our results in the theory of mechanics; and we would expect the final solutions to the problems of space and time only after extending the analysis to the concept of motion in electrodynamics. It therefore makes sense that Einstein *started with electrodynamics*; his famous first work (1905) bears the title "On the Electrodynamics of Moving Bodies," and this was the path to his "theory of special relativity" in which his critique of the concept of time may be found.

Motion in the context of electric current already led to strange conclusions. It was quickly discovered that we are not looking at a continuously flowing ideal liquid, but rather at the transport of very small elementary particles, tiny atoms of electricity; however, the mass of these "electrons," unlike the masses of other bodies, is not a fixed quantity but increases with velocity. We therefore do not have Newtonian "point-masses." Nevertheless, we also have relativity here. Movement of electrons will give rise to an induction effect in a neighboring wire that is at rest; but we reproduce this effect precisely if the electrons are at rest and the wire is moving. This is well known to every electrician; in principle it is irrelevant to a dynamo whether the anchor or the field rotates. Again, relativity applies in this context, even for rotation: it cannot be determined by electrical effects which is rotating and which is at rest. This should strike us as bizarre, since the concept of relativity comes from a kinematical notion in mechanics and initially was not related to electricity at all. But even with all of the differences, there are deep similarities with mechanics. In both cases, we are dealing with the transporting of entities stripped down to their essences, be it electrical or material; and while the movement of electricity is not directly observable because the individual particles are so small, still no one can doubt the justification of the mechanical ideas in the theory of electrons. We just arrived at a good explanation of all phenomena by understanding electrons to be physical particles that move according to the laws of mechanics and carry only an electrical charge with them. The only remarkable thing is that electricity is more generally relativistic than mechanics, since it was always extended to include accelerated motion.

One can only expect something entirely different by considering wave motion that does not displace the particles.

There is already a difference in mechanics between *wave motion* and physical displacement; the motion of a water wave, for example, is in a sense imaginary as the water itself does not move, it is only the *shape of the surface* of the water that changes. Nevertheless, we may still consider this sort of motion to be a displacement. This can be seen most easily by considering the so-called *theorem of addition* of velocities that we will derive for such a displacement. On a glass plate, represented in the figure with a solid double line, a beetle crawls from A to B in one minute. Meanwhile, we slowly move the glass plate to the side so that after the minute its location is now represented by the dotted line; the point B now lies at the point C. When viewed from the table upon which the glass plate lies, the beetle has not moved along AB, but rather along AC; one can imagine this to be the path of the beetle's shadow that the sun would cast through the glass plate onto the table. By velocity, we of course mean distance per unit time. With respect to the glass plate, the beetle's velocity $v = AB$, but with respect to the table it is $w = AC$ because this is the distance covered in one minute from the perspective of the table. Call g the velocity of the glass plate, we get from the *theorem of the addition of velocities* that $w = v + g$, where the addition symbol is understood in terms of the parallelogrammatic construction of the arrows. (This is called vector addition.)

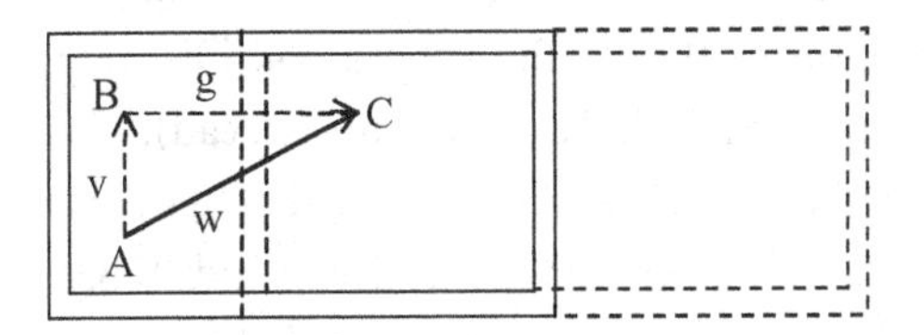

These considerations can be straightforwardly applied to the case of *wave motion*. Take, for example, a water wave with a velocity v and a ship moving with a velocity g, let w be the velocity of the wave as seen from the ship, it can be determined using the above method. Think of a stone thrown into water; it produces a circular wave spreading out in all directions. After a few seconds, it will reach a boat that is floating nearby. But if the boat is not at rest, but rather sailing in the direction of the stone, then it will reach the circular wave earlier. The boat's passenger can express this as follows: with respect to my boat, the wave has a greater velocity now than previously because it reached my boat sooner; the velocity of my boat has been *added* to the velocity of the wave. And the same must apply to the old view of electromagnetic waves. Since light waves are also waves, which we have known since Maxwell, the same addition theorem must therefore hold for light. Take, for example, a light ray with a velocity c with respect to an observer at rest, then the velocity as judged from a moving railway car depends upon the car's direction of motion – this had to be assumed before Einstein. Indeed, since the Earth moves at thirty kilometers per second in space, it had to be assumed that the speed of light on Earth is not the same in all directions. This depends upon which sort of medium propagates the light waves. It was assumed that electrical waves travel through a very thin, fine medium, the aether, in just the way that air propagates sound. According to the Copernican worldview, it was extremely unlikely that the substance that fills all of space would happen to conform to the motion of the Earth; but friction-like drag effects are highly unlikely given optical experiments (Fizeau). The American Michelson tested the question of the speed of light in a very exact experiment; contrary to all conjectures, he found (1883) that the speed of light on Earth has the same value in all directions.

The phenomenon of electromagnetic motion has led us to a consequence that could not be understood through the concepts of motion from kinematics and mechanics. Here our concepts run up against their limitations because they were derived from far too basic experiences that are no longer sufficient in the case of the complex phenomena of electric fields. We have found a *contradiction to the addition theorem.* Other ways of explaining these observed phenomena were searched for in vain as no one suspected that the error was to be found in the *concepts.* Einstein was the first to have the chutzpah to change the foundational concepts themselves and thereby was the first to attain insight into the true nature of motion.

We have just derived the theorem of addition in such a simple and obvious fashion that many readers may have passed over it; nevertheless, I ask for the reader's attention to this derivation. As with most foregone conclusions in the exact sciences: *the derivation is wrong.* It is *fallacious*; a precise examination of it exposes a presupposition concerning something that cannot possibly be known *a priori* and which we now know to be empirically refuted. We said that the beetle moves from *A* to *B* in one minute. How is this measured? We set up clocks on the *glass plate* at *A* and *B*. With respect to the *table*, we said, it moved from *A* to *C* in one minute; this is measured with clocks placed *on the table.* How do we know that the clocks *resting on the table* would also have shown one minute as having passed like *those on the glass plate*? Indeed, it will be objected that we naturally assume that the clocks *run at the same rate.* What do we mean by running at the same rate? I will assume that two clocks placed next to each other run at the same rate, for an hour, for days – it should be a clock that runs off of the tension of a wound spring so that it cannot be influenced by anything external. Does it follow from this that the clocks will display the same amount of time having passed if one clock was moved away and returned again to compare? This cannot be determined *a priori*; both results are equally possible. Einstein, however, considered: by first giving up the foundational principle that moving clocks run at the same rate as resting clocks, can't we then understand Michelson's experiment according to which the addition theorem does not hold for the speed of light? And thus the solution was possible. Michelson's experiment became obvious, and the addition theorem for velocity underwent a correction such that for *small* velocities it reduces to the old formula; for large velocities, however, it led to the strange result that whenever the speed of an object is added to the speed of light there is no increase in the speed of light. Electrodymanics is now completely relativized. In all inertial systems, i.e., systems free of gravitation, and in all directions the speed of light is the same. And furthermore: the new hypothesis explains not only the Michelson experiment, but also a wealth of other observations; the above-mentioned dependence of mass on velocity naturally derives from this (not only for electrons, but for all point masses), and the physicist understands with a glance so many phenomena that previous theories had not been able to explain at all!

This is the reality that the physicist has experienced – one can now perhaps understand with what reluctance the physicist has had to read the abundant pseudo-philosophical writings that have popped up like mushrooms since Einstein's rise to fame and claimed to irrefutably prove that Einstein's theory is (a) completely false, (b) utter nonsense, or (c) capable of being proved wrong by any third grader. One particularly deep critic has concluded, "If two events are simultaneous, it follows from the definition of simultaneity that they cannot be non-simultaneous." Another one recently wrote a letter to the Swedish Academy in Stockholm demanding the Nobel Prize because he had found the "miscalculation" in Einstein's theory. A brilliant, high-ranking medical officer sent a memo to German universities to inform them of his ultimate refutation of Einstein's theory because he had discovered the aberration that Einstein allegedly had not taken into account. Mixed into this chorus we find all kinds of off-key notes; one gentleman, who unfortunately was accompanied by a few physicists, felt the need to apply his political instincts to Einstein, and spoke of Einstein's "scientific dadaism" and "scientific larceny." With the dying down of popular interest in the theory of relativity, this clamor is dying down as well. But all the while, Einstein's theory has been put in place as a firm foundation for physical science.

The electrodynamical solution that was achieved through Einstein's special theory of relativity gave the problem of motion a completely new turn. It became obvious that there are some phenomena of motion for which the previous notion fails. Indeed the water wave in the old understanding was never a movement of anything material; however, the movement of energy in the water wave can be understood by considering it to be the *transverse* motion of individual water particles. We think of a water particle being struck so that it momentarily moves up perpendicularly above the surface of the water; this pulls along a neighboring particle, which pulls another one, and thus one after another the water particles are lifted upwards after the first one has already sunk back down again. This results in the picture of a laterally moving wave. For *light waves*, however, this understanding is not tenable. One may want to keep the name *aether* for empty space because space has the physical property of being transparent to light, but one can *never* think of this *aether* as having substance. It is missing the elementary property of substance, i.e., a particle can always be individually identified in its specific path. The explanation that was given for water waves is not applicable to aether waves. In the same way, one also cannot understand light as a physical substance hurled through space. We are faced with the obscurity of the existence of motion that does not result from the motion of material things; the concept of motion is profoundly more fundamental than that of substance. The mechanical concept of motion is not at all the foundational phenomenon that philosophical materialism contends – standing prior to it is the electromagnetic concept of motion from which the phenomenon of substance and the physics of motion can be derived through the methods of physics. We must keep in mind that all *understanding* in physics is nothing but a quest for analogies with day-to-day experiences; but we ought not expect that our commonplace concepts are always as they ought to be.

Natural science has traditionally taken the concept of *substance* to be perfectly clear and "motion" always meant as the movement of *substances*, but a strange inversion now occurs as a result of the theory of relativity: the investigation of the phenomenon of *motion* leads to a revision of the concept of *substance*. Only after this was accomplished could we readdress the problem of *mechanical* motion. We have already shown that this problem requires an analysis of *gravitation* because it requires *dynamical gravitational effects*. However, if one adds the new thought that there are no decisively demarcated bodies, but only fields with knot-like condensations (matter) that are only a continuously dispersed something in space that leaves no holes, then one thereby reaches a critique of *space*. *Space* is then nothing else but this *something*, and *geometry* is the *law governing its structural relations*. Space is no longer the great apartment building that houses all solid bodies, but rather the structural schema of the general fields from which the real world is built.[1]

But it is here that the physical problem merges with the philosophical. What then are space and time if they are only metrically defined through the field, i.e., the "real" in the world? Are our intuitive notions of straight line and plane meaningless? What then do we mean by knowledge if even our oldest forms of thought fail? The answer is not so easily found. I have set out my own position elsewhere.[2] But I advise anyone who is not willing to surrender old dogmas to avoid considering these matters any further.

[1] I have shown that only this view of geometry leads to sensible consequences in an earlier essay, "Einstein's Theory of Space," *Umschau* 26, June 1920. [See Chapter 2 in this collection.]

[2] H. Reichenbach, *Relativitaetstheorie und Erkenntnis Apriori.* Springer, 1920.

7

The Theory of Relativity and Absolute Transport Time

In a recently published article,[1] I have reported on an axiomatization of the relativistic space-time theory. In light of this, we may now test the possibility of *absolute* time by investigating which axioms are and are not compatible with it.

One possibility for defining the synchronization of clocks, so that the same simultaneity relation holds for all systems, arises from the transportation of clocks.[2] Two clocks that are brought into synchronization at the same place are to be called synchronized as well when one or the other is transported to a different place. This is a *definition* of simultaneity and is neither true nor false, but an arbitrary rule. For that reason it can be used in any

Translated from "*Relativitätstheorie und absolute Transportzeit*," *Zeitschrift für Physik*, vol. 9 (1922), pp. 111–17.

1 Phys. ZS. 22, p. 683, 1921. The detailed publication is in preparation; this is the announcement of a further result of current interest. [See Chapter 4 in this collection.]

2 This is not the only possibility.

case; but in order to be *univocal*, it must satisfy the following axiom:

Axiom A: Two clocks that are synchronized at one place are always synchronized when compared at the same place regardless of the paths of transport.

By "clock" we understand here a closed periodic system.[3] Whether axiom A is satisfied is a mere matter of fact and can be decided independently of the definition of simultaneity for distant places and independently of the theory of relativity. The theory maintains 'and it has received extremely indirect confirmation' that axiom A is false, and therefore it rejects absolute time; but to the present, no means had been found to directly test the axiom. For that reason the following question arises: From which axioms of the theory of relativity does it follow that axiom A is false?

We want, however, to further simplify the question, since we want to consider only the special theory of relativity. Axiom A includes accelerated motion of the clocks as well, and it is very possible that accelerated motion will result in a retardation of the clocks while uniform motion will not. The former can be accepted when the red shift is experimentally confirmed; however, for the case of uniform motion it remains undecided. Synchronization can be defined in the special theory of relativity just by clocks that are uniformly transported. Consider two distant clocks, U_1 and U_2, which are to be synchronized; consider a third clock U_3 to be transported uniformly from U_1 to U_2, such that it is set to be synchronous with U_1 when it passes by it and transfers the synchronization to U_2 instantaneously when U_3 and U_2 pass each

[3] The conceptual difficulties inherent in the notion "closed" will be addressed in a different submission. [See Chapter 7 in this collection.]

other. In this way an absolute time can be defined for the special theory of relativity; its univocality is fixed only by satisfying the following less stringent axiom:

Axiom B: The resulting synchronization is independent of the magnitude and direction of the uniform velocity of transport.

Here, however, a difficulty appears. So long as simultaneity is not defined for any coordinate system, uniform translation is also not defined. In at least *one* system, simultaneity must be defined in the fashion of Einstein; then uniform translation relative to this system is defined, and axiom B can be used. We want here to define the concept "uniform" in this way for use in the manner of axiom B. This is *not* an axiom, only a definition. It is *not* necessary that the resulting synchronization under transport in this system is identical with that of Einstein; it need only be univocal. At a later point we will add the restrictive assumption (axiom C) that this identity is valid for some given system. But even then, one is not also forced to use this distinguished system as the basis system for the definition of simultaneity.

Axiom B will be denied by both the special and general theories of relativity; therefore the above-mentioned question remains to be dealt with.

We handle it in the following way. In the previously mentioned report, the axioms are divided into *light axioms*, those that assert only the properties of light, and *matter axioms*, which assert the properties of rigid measuring rods and natural clocks. It is clear that axiom B can only enter into a contradiction with the *matter axioms*, since it asserts nothing about light. We will omit two of the matter axioms, specifically, VIII and X, and begin by showing that axiom B is compatible with the remaining axioms.

We will prove this by deriving transformation formulas for two uniformly moving systems K and K' from the combined axioms without internal contradiction. For this derivation, we can add further arbitrary definitions because the choice of the definitions will not influence the result; only the *form* of the transformation formulas depends upon it, but not their freedom from contradiction. In choosing suitable definitions we only simplify the calculations.

The first omitted axiom, VIII, says that the length $(= l_K^{K'})$ of a rod at rest in K measured in K' is equal to the length $(= l_{K'}^{K})$ of the same rod at rest in K' as measured in K; it therefore satisfies the relativization condition for length. The second omitted axiom, X, refers to the principle of the constancy of the speed of light. This principle says two different things: first that the speed of light is a constant of the system, and second that this constant becomes the same number in all systems when measured with the same measuring rods and clocks. Axiom X contains the second part of this claim; the first already follows from axioms I to VII and will therefore hold for successive ones. Accordingly we will omit definitions 9 and 10 and replace them with the following:

Definition a. The transmission of the unit of length occurs through the transporting of a rigid measuring rod.

Definition b. The transmission of the unit of time occurs through the transporting of a natural clock.

These are the suitably selected definitions. But we introduce one additional limitation into the derivation of the formulas. We add an axiom C. When we have shown that axioms B *and* C are compatible with axioms I through VII, and IX, it will show the independent compatibility of axiom B; the required proof is therefore not restricted. Over and above this, axiom C has a deeper justification. Axiom B alone does not guarantee the transitivity of synchronization by transport, although axiom A alone is sufficient to guarantee it. But the Einsteinian synchronization is transitive (this follows from the light axioms), so when we add axiom C to axiom B, transport synchronization also becomes transitive. One can hardly set out a theory of absolute time that does not possess transitivity.

Axiom C: There exists a system K in which synchronization by transport is identical to the Einsteinian synchronization in definition 5.

It can be shown that this identity does *not* hold for a different system. We nevertheless define synchronization in another such system K' according to Einstein (definition 5), thus not employing absolute time; the admissibility of this move follows from the remark made above. We look for only transformation formulas that do not *exclude* synchronization by transport.

Axioms I through VII, and XI; definitions 1 through 8, a and b; and axioms B and C produce the following transformation formulas according to a simple calculation:[4]

$$x = vt' + \frac{c}{c'}x', \qquad x' = \frac{x - vt}{\dfrac{c}{c'} - \dfrac{v^2}{c \cdot c'}}$$

$$t = t' + \frac{vx'}{c \cdot c'}, \qquad t' = \frac{t - \dfrac{v}{c^2}x}{1 - \dfrac{v^2}{c^2}} \tag{1}$$

[4] The derivation is omitted here because of limited space, it contains no additional problems. The transformations for the coordinates perpendicular to the direction of motion are also omitted; they are in general not the identity transformation.

c and c' are the speed of light in K and K'; since axiom X has been omitted, the two numbers are different. v is the velocity of K' as measured from K. The formulas (1) are free of contradiction, which completes the required proof.

It can be easily seen that the formulas (1) exclude the relativistic time dilation. Let $\tau_K^{K'}$ be the time elapsed on a clock at rest in K as measured in K' and $\tau_{K'}^{K}$ be the time elapsed on an identical clock at rest in K' as measured in K (i.e., the proper times of the compared clocks will be the same), so we have:

$$\tau_{K'}^{K} = \tau, \qquad \tau_K^{K'} = \frac{\tau}{1 - \dfrac{v^2}{c^2}},$$

therefore

$$\tau_{K'}^{K} \neq \tau_K^{K'}.$$

The relativity principle for clocks will therefore be violated by the transformation (1); but this is easy to see using assumptions B and C. Axiom C fixes a coordinate system as privileged; for that reason the inverse transformations of (1) are also in a form different from the direct transformations.

The question now arises whether or not at least one of the omitted axioms can be added to this combination of axioms. We begin by adding in X; i.e., we set $c = c'$. Then the equations (1) become:

$$\begin{aligned} x &= vt' + x', & x' &= \frac{x - vt}{1 - \dfrac{v^2}{c^2}} \\ t &= t' + \frac{v}{c^2}x', & t' &= \frac{t - \dfrac{v^2}{c^2}x}{1 - \dfrac{v^2}{c^2}} \end{aligned} \qquad (2)$$

These formulas are also free from contradiction. It therefore also follows that axioms I through VII, IX, X, B, and C are all compatible.

Here, however, the demand of relativization of spatial distance is no longer satisfied for spatial lines. From (2) we get:

$$l_K^{K'} = l, \qquad l_{K'}^{K} = l\left(1 - \frac{v^2}{c^2}\right),$$

$$l_K^{K'} \neq l_{K'}^{K}.$$

Similarly, therefore (because of the validity of VII) the identity transformation no longer holds for the axes perpendicular to the direction of motion.

We get another combination when, instead of X, we add VIII to the above combination. From (1) we are given:

$$l_K^{K'} = \frac{c'}{c} \cdot l, \qquad l_{K'}^{K} = l\left(\frac{c}{c'} - \frac{v^2}{c \cdot c'}\right)$$

with VIII, i.e.,

$$l_K^{K'} = l_{K'}^{K},$$

we arrive at the condition

$$c' = c^2 - v^2. \tag{3}$$

Substituting this into (1) yields

$$\begin{aligned} x &= vt' + \frac{x'}{\sqrt{1 - \dfrac{v^2}{c^2}}}, & x' &= \frac{x - vt}{\sqrt{1 - \dfrac{v^2}{c^2}}} \\ t &= t' + \frac{vx}{c^2\sqrt{1 - \dfrac{v^2}{c^2}}}, & t' &= \frac{t - \dfrac{v^2}{c^2}x}{1 - \dfrac{v^2}{c^2}} \end{aligned} \tag{4}$$

Since (3) is a possible stipulation for c', (4) is therefore also free from contradiction. It therefore also follows that axioms I through VII, VIII, IX, B, and C are all compatible.[5] The contradiction first arises when X is added because $c = c'$ is inconsistent with (3).

Accordingly, we obtain the following result. While the absolute time of transport is excluded from the complete set of the relativistic axioms, this exclusion is only due to axioms VIII and X. The time of transport is compatible with those axioms individually. Not even the principle of the constancy of the speed of light by itself excludes absolute synchronization by transport; even if the speed of light were the same number for all systems, synchronization by absolute transport would still be possible. The Michelson experiment (axiom VII) in particular cannot exclude it. Only the conjunction of the "light principle" and the principle of relativity make synchronization by transport impossible. It is known that these two principles work together: axiom X speaks only to the connection amongst clocks and measuring rods within the same system, and axiom VIII lays down the relationship between moving measuring rods. Only the conjunction of the two determines the transformation of the units and it is determined in such a way that denies absolute time of transport.

Axioms VIII and X are not directly confirmed yet; these axioms can only be argued for by analogy. It would therefore be important if the Einsteinian time delay were to be directly proven (transverse Doppler effect). Only then will synchronization by transport be empirically refuted.

What would it now mean if these observations turn out to be negative, i.e., if the time of transport were still possible in

[5] In this case, the coordinates perpendicular to the direction of motion transform according to the identity transformation.

violation of Einstein's assumption? One result is that there would exist a process to define simultaneity in a natural way so that it is known to be the same in all systems and can be called "absolute time." It would follow that *either* the Einsteinian light principle *or* the relativity principle for rigid rods must be held to be invalid. Yet it should not be believed that that the relativistic concept of time would be false.[6] As I have shown in the former report, this time can be defined purely in terms of the light geometry without using natural clocks and rigid measuring rods. Since definitions are arbitrary, the two times are equally justified. Independent of all empirical verification it can therefore be said: if there were an absolute time, it would no longer be absolute.

[6] Fr. Adler (Ortseit, Systemzeit, Zonenzeit, Wein 1920) advocates this view in this strange way. He uses synchronization by transport as a definition of absolute simultaneity relying also on assumption C. In this way, he derives formula (2) (p. 44), however, in a different relation. It is to Adler's credit that he has noted throughout that the assumption $c = c'$ is not *necessary*, especially not following the Michelson experiment. Certainly this is an obvious insight. Adler then commits a mistake when he assumes that $c = c'$ is *not possible*; so, as we have shown here, it does not contradict absolute synchronization by transport. It is an empirical question whether the assumption is valid; its logical possibility is derived throughout the given axiomatic account.

8

Reply to Anderson's Objections to the General Theory of Relativity

In number 5114 of volume 214 of this journal, Mr. *Anderson* has raised several objections to my reply to Mr. *Wulf*. I have only now become aware of them, so my response will seem belated. All the same, I do not want to forgo this response because it is important to clarify the basis of this incessant misunderstanding of the theory of relativity.

Anderson admits that the theory of relativity provides a contradiction-free explanation for the "relativistic" perspective of the carousel at rest and the stars rotating with large angular velocity. He believes, however, that the theory becomes flawed when the direction of motion reverses; he contends that it is a coincidence in the relativistic perspective that all stars reverse their direction of motion at the same time. Let me begin by saying that obviously the changes in the directions of motion are contained in the differential equations of the $g_{\mu\nu}$-fields. Further, it is not a matter of chance

Translated from "*Erwiderung auf Herrn Andersons Einwände gegen die allgemeine Relativitätstheorie*," *Astronomische Nachrichten*, vol. 213 (1922), pp. 374–5.

that the motions of the stars define the celestial axis as a privileged straight line, but rather it is well grounded in the distribution of stellar motions; the same conditions that, from the non-relativistic viewpoint, place the Earth at rest, also determine the distribution of the stars from the relativistic perspective after an appropriate transformation. Hence, this can in no way be called a coincidence. The reason *Anderson* responds in this way is that he is trying to account for the gravitational field with concepts from the *Newtonian* theory, and of course the *Einsteinian* notions would thereby lead to contradictions. This appears to be the most common sort of mistake made in judging the theory of relativity; naturally, consistency can only be expected within a theory when one limits oneself to the concepts of that theory. In physics, there are two different ways of measuring: absolute measurement and frame-dependent measurement; the first (e.g., the electrical load) is independent of the coordinate system, while the other (e.g., velocity) is only defined once the coordinate frame is fixed. Where gravitation is an absolute quantity for *Newton*, it is a coordinate-dependent quantity for *Einstein*; and when this is taken into account, the flaws Anderson believes he has uncovered immediately disappear. An example will illustrate this. Consider two ships sailing alongside one another with different velocities. The speed of ship *B* relative to that of ship *A* is a well-defined physical quantity. If *A* then changes its course, then we must also change the defined speed of *B*; but surely no one would call it a coincidence that the change of velocity of *B* always results whenever *A* changes its course. Velocity is, of course, a relative concept, it describes a relation between things. If we consider the "gravitational field" to be a relative concept – and this is what is intended by the general theory of relativity – then *Anderson's* so-called "coincidences" simply vanish.

The only objection that can therefore be raised against the theory of relativity is that it is impossible for the gravitational field to be interpreted as a relative concept. But precisely this objection is not allowed because the theory has shown that with this concept all observable phenomena can be accounted for, in fact it can better describe them than the absolute concept of gravitation. Further, the gravitational fields cease to be "spooky," but are as real as any other physical object and become, for example, experimentally demonstrable. Contrary to *Anderson*, we cannot look for the gravitational-rotational field in the vertical component of the Earth's field from Montblanc; but it has been experimentally demonstrated using the Coriolis force in the horizontal direction that is the most experimentally determinable.

I must therefore reiterate my original opinion that there is little sense in raising conceptual objections to the theory of relativity. These sorts of objections arise only for those who cannot accept the theory's system of concepts. It seems more important to break the theory down to its observable consequences and to search for measurable effects that can be used to test it. Astronomy in particular must play a crucial role in this because it will probably long remain the only science capable of measuring Einsteinian effects.

9

Review of Aloys Müller's *The Philosophical Problems with Einstein's Theory of Relativity*

Braunsweig, Fr. Vieweg & Son, 1922. VIII, 224 pages and 10 illustrations. 8°. Price, M. 7,50 stapled and M. 9,25 bound.

A comprehensive work that sets out the philosophical problems of the theory of relativity in detail would be a true enrichment of the literature. Unfortunately, *Müller's* volume contains mistakes. It distinguishes itself by committing these errors in a very dull and matter-of-fact way and not in the usual emotion-laden warlike fashion of most of Einstein's opponents – yet it is built on an entirely flawed understanding of the special theory of relativity that results in faulty conclusions. A necessary prerequisite for any philosophical critique of relativity is an analysis of the theory's factual basis and conceptual foundation; and *Müller's* efforts are frustrated by the inability to offer such an analysis. His first mistake concerns the concept of simultaneity. He claims that simultaneity can be *known* and does not need to be *defined.* He fails to grasp that there are two types of definitions: the definition

Translated from *Die Naturwissenschaften*, vol. 11, no. 2 (1923), pp. 30–1.

of concepts within a conceptual system and "coordinative definitions" which specify how given concepts acquire an empirical reality. Thereby, a conceptual definition is posited for a unit length; but in order to actually carry out measurements a coordinative definition must be put in place to fix that "this rod here" is a unit long. Similarly with simultaneity; the *concept* will not be dependent upon physical realities, but in order to determine what events we call simultaneous we first need the Einstein convention and help from light signals. This convention is *arbitrary* – any other of convention would likewise be arbitrary. This initial misunderstanding is perhaps the basis for all of *Müller*'s mistaken claims about the space-time of the theory of special relativity. For example, he claims that the principle of the constancy of the speed of light is a *mathematical fiction* that has nothing to do with real things. He overlooks the fact that, besides statements about light, the theory of relativity also makes *factual assertions* about the way measuring rods and clocks adjust themselves in accordance with the constancy of the speed of light; and thus the constancy principle becomes an empirical proposition. Further significant errors can be found, e.g., that two observers moving with respect to one another will read one and the same clock differently – this error originates again in *Müller*'s complete misunderstanding of Einstein's explanation of simultaneity.

These are only a few of the notable points; but they are enough to show that a philosophical assessment of the theory on these grounds is impossible. The general theory would also suffer from the same misinterpretation as the special theory. A few of the objections that were raised against the explanation of gravitation are extremely superficial. E.g., *Müller* demands that the relativity theorist provide astronomical confirmations that could be explained by *no* other theory as evidence for the theory of relativity – an entirely impossible demand that could be satisfied by no other physical theory. There are in principle an infinite number of explanations for any fact. The *choice* between them is the real problem and can only be solved within the framework of a theory of induction.

I do not dispute that this book is a serious work of philosophy. If it had appeared 2 or 3 years earlier, it would have been at least considered a step towards clarifying the problems of relativity. But since all of these questions have already been clarified and the investigation is in the process of moving to yet deeper levels, the book comes to this difficult project too late.

10

The Philosophical Significance of the Theory of Relativity

Introduction

Although there is still resistance to the theory of relativity, it should be pointed out that this resistance is founded upon *conceptual* objections. It is beyond dispute that the theory is *physically useful*, that its assertions are well verified by observable phenomena. What opponents of the theory find problematic are the ideas upon which the theory is founded. On the other hand, it is precisely these ideas that the defenders of the theory hold to be its greatest achievement and in which they claim to find the true significance of Einstein's work. It is therefore of interest to study the formation of these ideas, their content and their significance.

We begin by rejecting two ineffective objections. The theory has been criticized for contradicting ordinary common sense. It is necessary to concede that this is true, but we refuse to see this as a *criticism* because a theory like this, which provides an analysis

Translated from "*La signification philosophique de la théorie de la relativité*," *Revue Philosophique de la France et de l'Étranger*, vol. 94 (1922), pp. 5–61.

of the most profound abstract ideas, will necessarily contradict certain naïve intuitions from everyday life. We do not discount the value which is attached to this simplicity of understanding, but a mentality adapted to the practical needs of existence (and is ordinary common sense anything else?) cannot be required to possess the proper critical faculty of a theory of knowledge: "The chisels and hammers are fine for working a piece of wood, but to engrave you need an engraver's needle," these words in Kant should not be far from view whenever one wants to contradict the theory of relativity with elementary objections.

Another criticism asserts that the theory of relativity only possesses a *mathematical* significance, that it is a sort of game based on fictitious measurements. A well-verified theory that is capable of deducing so many phenomena is necessarily given a significance which exceeds the domain of pure *thought*; it is a manifestation of a connection with *nature*. Indeed, the problem here is that it is difficult to have a direct acquaintance with this connection, but, since it exists, the evolution of the concepts of the theory must be sensible. When they say that Einstein's notion of time is nothing but a calculational fiction, they are entirely forgetting that this calculational fiction allows for the deduction of very real consequences. It must be the case that either all physical concepts are calculational fictions or that affording this designation to Einstein's notion of time is based on a misunderstanding. That this notion of time is not immediately present to intuitive representations in the way that the older physical concepts are is an objection not to this concept, but to our faculty of representation. Furthermore, it should be noted that Einstein's notion of time becomes perfectly intuitive with a little work with the relativistic concepts. If one is careful to come to understand the theory by thinking abstractly, the reproach that it is counterintuitive is unsupported – indeed, this reproach could be applied to the creation of every new thought and would undermine all of them. It is important to follow the concepts by which the theory finds its way step by step and to level criticisms at the theory from the *same* intellectual path as was used in the creation of the theory. This work derives from such an attitude; indeed, to renounce this method would achieve nothing but to promote traditional representations to a status of absolute predominance.

We will have mistaken everything that is found in the work of Einstein if it is taken only as constructing a *physical* theory, since he was ever cognizant that the introduction of a new theory always involves *philosophical* discovery. The initial problem of the theory of special relativity (the contradiction between two optics experiments) is a problem of comprehension, not physical discovery. These two optics experiments[1] were only contradictory in the sense that we *understood* each through different theories. The *physical* discovery was achieved through the performing of these experiments, it was only the *logical* discovery of their co-intelligibility that remained. It is true that all physical theories are *intellectual* products because they are designed exclusively to establish a logical connection amongst observed facts. But in this case it seems that all the usual intellectual methods have failed. Lorentz established his theory through the framework of classical methods, but his theory itself led to a new and unintelligible result, the contraction of rigid rods. While the confirmation of Lorentz's theory does not contradict the concept of causality (because properly speaking it has a causal explanation), its unintelligibility lies in the fact that all the actions arising from the aether are *governed by quantitative laws which only make use of statements concerning*

[1] The experiments of Michelson and Fizeau.

motion relative to the aether. Physics cannot consider such an effect as fortuitous; it must search for an explanation that will render understandable the existence of *universally unverifiable* facts. This is the beauty of Einstein's solution, it makes this unverifiability understandable via the principle of relativity, i.e., by giving up the material structure of the aether that fixes a privileged state of absolute rest. Ehrenfest perfectly summed up the innermost sense of Einstein's theory when he joined together the following three statements:[2]

I. Light sources send us light signals as independent phenomena through empty space.
II. The measured velocities of light rays from a source moving towards us and a source at rest would be observed to be the same.
III. We declare that we are satisfied with the combination of these two statements.

But the combination of these two statements only becomes truly understandable if one admits the Einsteinian definition of simultaneity. In fact, the measurement of velocity presupposes the definition of simultaneity, and if this does not mean the same thing for observers in different states of motion, (II) is not in contradiction with (I), that is to say that the constancy of the speed of light (II) is not in contradiction with the principle of relativity (I) because (I) implies the negation of a material aether. The contradiction of the optics experiments is not resolvable through the framework of classical concepts, but requires the philosophical analysis of the notions of space and time necessary for the construction of the theory of relativity.

2 *Zur Krise der Lichtæther-hypothese,* Berlin, J. Springer.

Is not the "principle of the constancy of the velocity of light" lacking in sense? Can something incomprehensible be clarified through the positing of an incomprehensible principle? Certainly *not*, and Einstein is aware of this fact. His theory does not make a triumphant march against the world of thought. But it is an absolute mistake to object to the theory of relativity on the grounds that it is unintelligible. It is as intelligible as all others and in no way contradicts reason; furthermore a deeper examination will show that it satisfies our rational desire to comprehend the universe better than classical physics does. The contradiction between the two optics experiments rests, to be sure (and this is Einstein's most important discovery), on a hypothesis that we have always implicitly made and yet is only a limitation of our thoughts, a prejudice. A detailed analysis must follow concerning the thought: Why do we find it incomprehensible that a light pulse will have the *same speed* in two systems in relative motion? The response is far from being immediate, and all that ceases to be confirmed is the fact that it is incomprehensible to try to find the logical basis of its confirmation. To find the response, we need to formulate the problem in a simpler and more concrete fashion. Suppose that a light signal is emitted at a point A at time t; it will propagate in all directions, and we speak of "spherical waves" centered at A. Let a moving point A' be coincident with A at t. As time passes, it will move away from the center of the spherical waves. Will not the propagation of the light therefore fail to maintain its form as a spherical wave? Yet Einstein claims that for both A' and A, the surface of the wave will be that of a sphere. Here is the problem.

What is the surface of the wave? It is neither a material configuration nor something that one can contain: it is a mathematical structure that we *impose* on a region of space filled with light.

In the neighborhood of A an electrical force arises; at a point B_1, for example, the electric force (the "electric vector") has at time t_1 ($t_1 > t$) its maximum value. It changes at a later time, the force at B_1 becomes zero; later still it becomes again equal to what it was initially, but with the opposite sign: the vector at B_1 is oscillating. This phenomenon occurs at every point in the neighborhood if a lamp is lit at A and it is in the space that we term *illuminated.* Consider a point B_2 in the neighborhood of B_1 whose electric vector is also at its maximum value at t_1 and both have the same relative motion along the lines connecting AB_1 and AB_2. At this point the electric force has the "same phase" as at B_1. If we move gradually, we find that all of the points B with the same phase are the same distance from A, they are therefore on a sphere with center A: it is this which we call the "wave surface." Note that in the definition of this surface we introduce an important element: it is the geometric notion of equal phases *at the same time.* Without this last restriction we cannot posit our definition.

Consider, for example, not B_1 at time t_1, but a point C which has the same phase as B_1 at t_1 at a certain time *later*: we are no longer on a sphere. The shape of the surface of the wave is thus only determined once simultaneity has been defined, a hypothesis that is now exposed as a result of thinking that the wave is not spherical at A'. We have, in effect, accepted implicitly that all of the points B_1, B_2, etc. which form the surface of a wave for A also form the surface for A', although A is certainly not the center of the surfaces of the wave. But we have come to see that the shape of the surface of the wave is only defined via simultaneity: our implicit hypothesis is that simultaneity in the system A' has the same meaning as in the system A which is at rest. When we give up this hypothesis, the contradiction disappears. For A', the surface of the wave is not determined by the points B_1, B_2, etc. but by the points B_1, *C*, *D*, which according to *A* are on different surfaces. So nothing prevents these points from forming the surface of a wave at A', since this surface is only a structure imposed by definition on an illuminated region of space. The assertion that light always travels with the same velocity in systems in relative motion – and it is in this way that we use the term "spherical waves" – is therefore perfectly intelligible if we allow simultaneity to differ in different systems.

The philosophical problem of time rests on this point. Should we eliminate the notion of absolute simultaneity? One thing is clear: *if* we eliminate it, the Michelson experiment ceases to be incomprehensible, light *can* have the same velocity in the two systems. The physical facts can be brought into perfect harmony *if* we surrender this hypothesis. *But can we surrender it?* I believe that we should return to the question "Why not?" What is it that prevents us from adopting as our basic concepts those which best fit the facts? This is the question we will now consider.

Here is one objection. The relativity of simultaneity constitutes a *logical contradiction* because to say that the event E_1 is simultaneous with the event E_2 is just to say that they do not take place at different times, and if Einstein considers that latter to be every bit as justified as the former, we have a pure paradox.

There are philosophers who take this reason to be sufficient to prove that Einstein's theory of time is false. They are justified if Einstein's deductions actually entail a contradiction with logic, with the axiom *A* is *A*, because the foundation of logic cannot be weakened by any physical theory. But these philosophers are completely mistaken in their *interpretation* of the paradox discussed above. What we are led to, in fact, is a contradiction between logic, which we take as established, and precisely that which is in question, that is, simultaneity is an absolute concept. If simultaneity is a relative concept, the statement "E_1 is simultaneous with E_2" contradicts the statement "E_1 is not simultaneous with E_2" as little as the statement "Havre is to the left of Paris" and "Havre is to the right of Paris" contradict each other. The judgment depends completely on one's point of view. If one is to be precise on this point, all arbitrariness must be removed. When one says, for example, "for a given system, E_1 is simultaneous with E_2," one can no longer turn this assertion around. The problem therefore reduces to a question of whether "simultaneity" is a relative concept. But there are not deductions of pure logic that can resolve the issue *logically*. The two concepts are consistent, they do not lead to a contradiction with themselves.

The objection can be presented in a more sophisticated fashion. To add signifies joining one object to another. Adding the speed of light c to the speed v of the system must have a result that is larger, but the theory of relativity states that

$$c + v = c.$$

Is this not nonsense?

This equation is, in fact, nonsense. But the addition of velocities is a *physical* process, and it is impossible to show that it is represented by this equation. Consider a system in which the speed of light has been measured and is found to equal c and a second system that is moving relative to it at a velocity v. What will be the speed of the same ray of light if it is measured with the clocks and rods transported in the moving frame? This is the real problem of the addition of velocities. Its solution depends entirely upon the behavior of the clocks and rods. It is an addition of velocities, but only in a symbolic sense; *algebraic* addition is only a special case of this *symbolic* addition, and it is a matter of fact to determine what mathematical operation we shall use to represent this physical equation. The equation would better be written as:

$$c(+)v = c$$

where the symbol (+) designates the *symbolic* addition.[3] Our case of *symbolic* addition can also be formulated mathematically (that is to say reduced to algebraic addition). This is a transformation that has become known, thanks to Einstein, as the theorem of addition:

$$\frac{u+v}{1+\frac{uv}{c^2}} = w.$$

This formula is the mathematical explanation of the symbol (+); it gives, as one can clearly see, $w = c$ for $u = c$, that is to say, the result demanded by the physics.

One can therefore say here what we said above. The contradiction only exists if one presupposes that which the theory of relativity rightly questions. If the addition of velocities is an algebraic addition, the equation of Einstein is a contradiction, but only *in this case.* Therefore, logical reasoning does not contradict Einsteinian simultaneity.

We now examine another objection: while relativistic time does not contain a logical contradiction, it contradicts intuitions essential to reason. It is evident that two simultaneous events cannot be non-simultaneous. According to this objection, there exists above our faculty of logical deduction a particular power of reason that produces particular prescriptions in the matters of simultaneity. Some along with Kant call this "pure intuition" or

[3] Compare this with the chemical equation: 2 liters of hydrogen + 1 liter of oxygen = 2 liters of water vapor, which is also a contradiction if the symbol + is interpreted in this way.

an "*a priori* faculty"; others call it the "phenomenological experience." What justifies this? One must agree that there exists a sort of psychological need that forces us to impose some sense of simultaneity. If two men knock at my window, I can very well hear if the two noises are simultaneous or not. The judgment "simultaneous" is contained within the perception itself. Furthermore, this judgment is a necessary function of all perceptions without which they would be worthless. It seems, therefore, that there would be an "*a priori* faculty."

But let us examine this mode of reasoning. Are the noises from the two knocks at my window *really* simultaneous? The noises, that is to say, the two perceptions of the noises, reach my ear clearly and distinctly together. Once we distinguish between these perceptions and their *physical causes*, i.e., the knocks at the window, we observe that our *a priori* faculty teaches us nothing. It takes little time for the sound to travel from the window to my ear. For the needs of everyday life, we can consider this time to be zero and we also say that the knocks themselves were simultaneous because the perceptions were. But for exact measurement, we can no longer use this approximation, and it is for this reason that physicists invented very precise devices that allow for the measurement of such small time intervals. It is also the case that these differences become large enough that, in everyday life, we can no longer neglect them, for example, when we hear the noise of a distant cannon.

We have come then to recognize an important fact. It is important to distinguish between "simultaneous at a single place" and "simultaneous at two different places." The first notion of simultaneity alone is given immediately by the content of perception. The arrival of the signal at my ear is an immediate simultaneity of this sort. On the other hand, if the events which produced them are at different points in space, the immediate judgment of their simultaneity is flawed; it must be that from the origin (the cannon), the signal (the sound) was sent to a second point and only there is the immediate simultaneity produced. This judgment only concerns the *arrival* of the signal compared with a local event, for example, the position of the hands on my watch.

"Simultaneity at a point" can therefore be directly recognized as the result of an immanent faculty, while "simultaneity at two points" can only be inferred. It must be denied that simultaneity is a *psychological* matter as soon as it becomes a question of distant spatial points. We have come then to the problem of simultaneity that is posed in *epistemology*: How can we move from simultaneity at a single place, i.e., "coincidence," to simultaneity at distant points which cannot be experienced? It is in posing this sort of question that we arrive at our goal because we have explicitly stated the problem of simultaneity in a form appropriate to the theory of knowledge.

To help with this argument, we need to know a physical law, for example, the law of the speed of sound. But how can I measure the speed of sound? I must compare the time of departure of the signal from a point P_1 with the time of arrival at a point P_2 and divide the difference by the distance P_1P_2. But to measure these times, I need to know in advance the position of the hands of the clock at P_1 that is simultaneous with that at P_2, that is to say, the clocks at P_1 and P_2 have been synchronized. We see that the analysis of the problem has led us into a vicious circle: to measure the velocity of the signal I need to know what is simultaneous, and to determine simultaneity I need to know the velocity of the signal.

We have arrived at a surprising result. Simultaneity at distant points is only determined with the help of physics. This leads to a vicious circle. Is it possible to escape from it?

The path to the solution, Einstein found, is very simple: *simultaneity at distant points is the result of an entirely arbitrary choice.* It can only be *defined*, not *observed*. The set of all physical experiences cannot fix a particular notion of simultaneity: not without a previously set definition.

The recognition of this fact is of extreme importance. Although Einstein had already presented this in the *special theory of relativity*, it was not completely established because he also posits a special rule for synchronizing clocks. It follows that for clocks to be synchronized in this way, the speed of light must everywhere be the same in all directions. This is the meaning of Einstein's definition:

$$t_2 = \frac{t_1 + t_3}{2}$$

(t_1 and t_3 are the time of departure and time of arrival at P_1, t_2 is the time of arrival at P_2). This is an extremely advantageous definition, and below we discuss why. But this definition is not *necessary*, for example, one could also define:

$$t_2 = \frac{t_1 + t_3}{3}.$$

Now the speed of light ceases to be a constant, so it is no longer the same in opposite directions. This is not incorrect, just inconvenient. Other values for all physical magnitudes will be found, and we will end up with a perfectly consistent system of physics that allows for a complete and univocal description of the universe. These ideas were integral to Einstein's development of the structure of the general theory of relativity.

The true solution to the problem of simultaneity therefore is beyond doubt. If physics leaves room for a certain arbitrariness in the determination of times, it must be that every determination is justified in the same way and permits the construction of a system of natural laws. *From the viewpoint of epistemology*, this is the meaning of the statement: "*there is no absolute time.*"

Yet we will pose this question: Ultimately, could there be a physical means of distinguishing one definition of simultaneity from the others? A selection of this type will consist of choosing that determination of simultaneity in which natural laws take their simplest form. Einstein responds in this way: although the numbers which serve to measure the particular physical properties (force, field, etc.) can be different depending upon the definition of simultaneity, the *form of the laws of nature*, that is to say the relations between the measurements, is always the same. This is the claim that constitutes the principle of general covariance. It is therefore impossible to characterize a particular definition of simultaneity in a privileged manner by means of a physical system. This is the *physical* meaning of the statement: "there is no absolute time." This statement is the more restrictive of the two, adding something beyond the *epistemological* statement.

Yet we still must supplement this assertion. General covariance is only valid, according to Einstein, if the metric of space and time are defined in a determinate manner. One may, in fact, measure spatial and temporal intervals in an arbitrary fashion, but corrective forces (the gravitational potentials $g_{\mu\nu}$) must always be introduced such that for all infinitesimal regions measurement results remain the same as those which were obtained with rigid rods and natural clocks. This condition is indispensable because the form of the laws of nature must remain invariant under all definitions of simultaneity. It follows from the physical statement

"There is no absolute time" that there is a relation between simultaneity and the functioning of natural clocks. We will discuss this relation below.

Precisely because the theory of relativity considers every definition of time to be admissible, we have the power to choose a certain definition of time and to demand that all observers attend to this time. For example, an observer in motion can define time by means of that which an observer at rest would measure; then the principle of the constancy of the speed of light ceases to be valid for him, and yet he is capable of unambiguously describing all possible phenomena. But a time of this sort will not be called "absolute" in this sense because any other definitions can give the same result.[4]

The problem of absolute time consists in finding a precept that *when applied in the same manner in all coordinates systems, always yields the same definition of simultaneity.* Einstein's elimination of absolute time correctly implies that no such concept exists. This is a most important point. It could be true or false on physical grounds, but the criticisms of it belong to experimental physics. Consider two examples that confirm this claim.

First, we could try to define absolute time by the transport of clocks. Two clocks are brought together and synchronized. One is then transported to a point distant from the other. It can be claimed *by definition* that the clocks remain synchronized. But if this definition is to be univocal, the following axiom must be considered to hold: "Two clocks synchronized with each other when placed together are always synchronized if, after ordinary

[4] As an example of an empty absolute time of this sort we can cite time in Lorentz's theory that only differs from Einsteinian time in that there is an *arbitrary* system of coordinates to which the title "absolute" is applied.

transport along different paths, they are compared again and found to be synchronized."

The theory of relativity holds this claim to be false.[5] This is clearly a matter of fact, and the negation of absolute time rests on an experimental basis. Both the axiom itself and its contrary are *conceivable*, so it is only from reality that we can learn which is true.

Absolute time can also be defined if infinite signal velocities are allowed. The time of arrival of a signal could then coincide with its time of departure. But the theory of relativity teaches that there is no speed higher than that of light. This is why this new definition of absolute time is also not successful. Here again we have an empirical claim, the possibility of superluminal speeds *and* their non-existence in nature are both conceivable propositions. It is a question of fact. But experiments with electrons show that for increasing speed the energy must increase more rapidly following the square law and *for material particles moving with superluminal speed* this is already infinite. It is therefore impossible to attain speeds faster than light. It is a matter of empirical fact that is moreover perfectly conceivable. Here again absolute time is rejected for experimental reasons.

We now discuss why we have drawn the distinction between absolute time *in the physical sense* and in the *epistemological sense.* The physical statement is empirical in nature, i.e., it is possible for it to be true or false, but the epistemological statement does not depend upon this alternative. Even if the physical claims of the

[5] Cf. Reichenbach, "Relativitätstheorie Hund absolute Transportzeit," *Zeitsch. f. Phys.*, vol. IX, p. 111, 1922. [Translator's note: this is a typographical error in the original, the title of Reichenbach's dogged defense of the relativistic notion of time should read "Relativitätstheorie und absolute Transportzeit." See Chapter 7 in this collection.]

theory of relativity are false, if, for example, the above-mentioned axiom for transported clocks were to be confirmed, the epistemological claims would remain exact. To be sure, there is an empirically privileged time and it can be called absolute, but all other definitions are also permitted and can support an exact physical system. The principle of relativity then ceases to be correct because not only the measurements, but also the forms of the laws of nature will differ in the two systems. The theory of relativity therefore makes two claims about absolute time:

1. There is no absolute time.
2. If there were absolute time, it would not be absolute.

We now come back to the question posed in this section, viz., is the relativity of simultaneity admissible? We can now respond, not only is it admissible but it is *required.* When it is a matter of distant points, it is not possible to cast simultaneity as a psychological phenomenon, hence the analysis of the physics leads to two different senses in which simultaneity is relative. The analysis reveals that relativity is at the same time both logically necessary and a matter of fact. The philosopher must import this result into epistemology.

I. The Special Role of Optical Phenomena in Relation to the Characteristics of the Metric*

Although the solution of the problem of time consisted in rendering time completely relative, it is only in the general theory of relativity that this solution is fully realized. In the special theory, a special time with privileged physical properties is defined,

* Translator's note: this section is incorrectly labeled II in the original.

although the deviations from the absolute time of classical physics are fairly small. We do the same thing with the measurement of space which is also carried out in a particularly simple manner using rigid rods. We must examine these questions below because the metric of "natural coordinates," which is yet to be defined, also serves as the foundation for the general theory.

First, by employing natural clocks, a certain course of time is distinguished as *uniform*. We consider a natural clock to be an isolated periodic system.[6] The rotation of the Earth is an example of a clock of this sort for astronomers. Revolving electrons provide another example. There is a very old hypothesis in physics, a hypothesis admitted by Einstein, that there is a privileged position for the measurement of time using systems that are considered to have periods of *equal intervals*. This hypothesis contains the first statement that all of these systems conduct the *same* time, which is by no means evident. There is, to the contrary, a perfectly conceivable physical hypothesis that holds the opposite. If two clocks of this type are placed side by side and coincide at the beginning and end of their first period, do we know that they will coincide at the second period and all others as well? It is perfectly possible that it would be otherwise; one cannot make the objection that for one of the clocks the second period would have a different duration than the first, because there is no means of comparing consecutive periods of the same clock. To say that the second period is equal to the first is a *definition*, but this definition leads to a permanent means of measuring the intervals of the two clocks; it is a hypothesis. Einstein employs this hypothesis in establishing the measure of time in gravitation-free regions.

[6] The difficulties contained in this word "isolated" are elucidated in the following section.

The measure of space is based upon the behavior of rigid rods: here again we make a hypothesis analogous to that of the clocks, it is the hypothesis that two rigid rods which are the same length at one place will always remain of the same length when transported to different places along different paths in space. It is only because this hypothesis is well verified according to appearance that it is possible to make a univocal measurement of length with rigid rods. A second hypothesis can be added alongside the first, a hypothesis which belongs as well to classical physics, namely: that the geometry obtained in this way is Euclidean when space is free of gravitation. Being Euclidean can also be conceived of as a *definition* of space free of gravitation. This part of the hypothesis therefore maintains that *there exist* such regions and that they can be accounted for with any desired degree of exactitude by considering small enough areas. It is only after the measure of uniform space and time is defined that one arrives at a definition of simultaneity. This definition itself has already been shown to be arbitrary and was chosen by Einstein in the special theory in a fashion that gains the greatest advantage, and this requires us to explore this point more deeply.

Suppose that we want to synchronize all of the clocks in a system, we must start with a central clock *A* and adjust each of the other clocks according to Einstein's formula:

$$t_2 = \frac{t_1 + t_3}{2}.$$

Let *B* be another ordinary clock, *B* is therefore synchronized with *A* according to this rule. But this does not show that inversely *A* is synchronized with *B*, this is a conclusion that does not follow logically. It is, in fact, a question of a new physical assertion. If I send a signal in the opposite way from *B* to *A* which returns to *B*, it will not have been at *A* at the time indicated by the above method applied at *B*. This is not obvious, an example to the contrary can be constructed: if *B* is in motion relative to *A*, I can synchronize *B* with *A* by using the above formula, but one is easily persuaded by the calculation that inversely *A* is not synchronized with *B*. We follow Russell in designating synchronization to be a *symmetric relation*. The symmetry of synchronization is not given by the definition, it results from the physical properties of light that we call for short "the uniformity of the propagation of light" (this statement can be rigorously formulated as an axiom).[7]

It is less evident that synchronization is also transitive (in the logical sense of Russell). If *A* is synchronized with *B* and *B* is synchronized with *C*, will *A* be synchronized with *C*? The fact that Einsteinian time satisfies this condition comes about through another property of light that we call "circular symmetry" (knowing that the time of travel of light along a triangular path is the same in both directions).

These two properties, symmetry and transitivity, constitute a great advantage for Einsteinian time. We do not mean to say that classical time does not possess these properties; to the contrary, they were always presupposed as intuitive. The fact that the special theory of relativity also admits these properties means that its doctrine of time is not entirely different from the classical notion. The essential advance of the theory of relativity is only that it renders these properties and their observational basis completely clear. It is an indication of Einstein's philosophical depth that in his first

[7] Cf. for this section Reichenbach, "Bericht über eine Axiomatik der Einsteinschen Raum-Zeit-Lehre," *Phys. Zeitsch.*, vol. XXII, p. 683, 1921 [Chapter 4 in this collection].

work[8] he clearly explicates this correlation. Let us now examine what follows from the privileged properties of Einsteinian time.

If a light signal is emitted at a point P_1 at time t_1 and arrives at P_2 at time t_2, precisely these properties demand that it always be true that $t_2 > t_1$ because the properties in question require the clock at P_1 must therefore be synchronized with that at P_2 regardless of how all the clocks in the system were synchronized, viz., independently of the order in which they were synchronized. We have arrived at the result that a light signal along its path always receives indications of time that force it to run in the *positive* temporal direction. If one sends a signal from P_1 which moves *slower than* light it will be received at P_2 at time t_2' which for very good reason will be such that $t_2' > t_2$. All signals or material points which move with a speed slower than that of light will move in a positive temporal direction. It can again be said that a clock that moves with a material point displays the same direction for the course of time as other clocks that pass the point. Now, it is a well-established theorem of physics that the speed of light is the fastest speed existing in nature. It can therefore be said that according to Einstein's notion of time, all signals or material points move in such a fashion that the direction of their proper times coincide with that of the time of the system.

This is a great advantage because it satisfies an essential requirement of the principle of causality. The principle demands that the cause precede the effect in time. If one calls a stationary material point at P_1 the cause of a sound arriving at P_2, the facts that are explained demonstrate that the privileged positive direction in time is acknowledged as positive throughout the entire system, and we note that the particular characteristics of time in the special

[8] *Ann. der Phys.*, p. 891, 1905.

theory of relativity entail the extremely important consequence that the principle of causality remains respected.

It is very important to make clear that the principle of causality itself is not absolutely necessary. A physics that violates it is not necessarily false; to the contrary, because of general covariance, physical laws maintain the same form as they would in a causal physics. It may appear paradoxical to speak of laws of physics as non-causal, but it must be noted that being subject to a natural law is more general than governance by the law of causality. It is only a very special type of law that possesses diverse, well-determined properties that are precisely like the privilege of a positive direction of time.

We remark that these unique characteristic properties of Einstein's time are related to the use of light as signal. It is therefore indicative of the investigation that the conditions will change if a different signal is used, for example, sound. Opposition to the theory of relativity is often founded on these grounds, that sound could just as well serve as a signal and that, *ad absurdum*, a theory of relativity could be developed requiring the impossibility of a speed greater than that of sound. This line begs for a response. A definition of simultaneity based upon sound signals according to the formula $t_2 = (t_1 + t_2)/2$ is not *false* because such a definition is arbitrary. Furthermore, it leads to a symmetric and transitive synchronization. But the inconvenience of such a notion of time resides in the fact that it would violate the principle of causality. If we insist on respecting the principle of causality, then it must follow that we cannot conclude that there are no speeds greater than that of sound. It must be said conversely that it is an experimental fact that light possesses a limiting velocity, and it follows that time is defined by optical signals that do not violate causality. As sound does not represent a limiting velocity, this conclusion is not valid for sound; it is precisely because of this that sound is not convenient as a signal for time in applying the formula of Einstein:

$$t_2 = \frac{t_1 + t_3}{2}.$$

Here the unique position of light is clearly distinguished.

In this respect it is important for the investigation that we state the proper formulation of the principle of the constancy of the speed of light. This theorem has been greatly contested, taken by some as impossible, by others as an indemonstrable but indispensable postulate, by yet others as an experimental fact, and by still others who, because of its strangeness, consider it to be empirically exact and prefer to eliminate it from the principles of the theory, deriving it instead as a consequence of the other hypotheses. This multiplicity of opinions is founded upon reality: a more exact analysis will show, in fact, that the proposition discussed is not a single proposition, but a combination of several hypotheses and a definition. The critics do not see that their objections are but elements of this combination and so obtain contradictory results.

The definition that is used here is that of Einsteinian simultaneity; it makes no sense to question its validity, because it is arbitrary. As this definition is indispensable for the positing of the constancy of the speed of light, this part of the principle is free from all observational criticism. Its indemonstrability is not open to criticism because there is nothing here which is subject to demonstration. Hence, it must not be believed that the definition of simultaneity is *sufficient* to obtain the statement; other hypotheses must also be admitted in order to derive the constancy of the speed of light from a definition. These hypotheses can all be formulated independently of the definition of simultaneity. For example: Let AB and AC be two lengths which are equal when measured with rigid rods, then the time it takes for a light signal emitted at A and reflected at B to traverse the path ABA is equal to the time taken for the corresponding signal ACA. The solution of this problem depends only on the point A, independent of any definition of simultaneity. It is incontestably a question of experience, and it exactly reproduces the content of the Michelson experiment. With the help of such hypotheses and the definition of simultaneity one can derive the first part of the "light principle," i.e., the statement: "in a gravitation-free system of natural coordinates the speed of light is constant."

The second part of the light principle is: "The characteristic constant c has the same numerical value in all systems if all measurements are taken with the same rigid rods and the same clocks." This therefore states a relation between the uniformity of transported clocks and the uniformity of length of transported rods.[9] This statement is consequently experimental in nature. Unfortunately, this cannot be empirically confirmed because the measurements are not yet sufficiently precise. It is a question of measuring the motion of electrons and in particular the important *transverse Doppler effect* of Einstein.

It can therefore be seen that Einstein's light principle is nothing strange and is not impossible to represent. It is a proposition like all other propositions in physics. It is made up, like all other physical propositions, of hypotheses and definitions. The fact that the speed of light plays a particular role comes from the fact that it is a limiting velocity.

[9] For the connection between this relation and the time of absolute transport, see Reichenbach, *Zeitsch. f. Phys.*, vol. IX, p. 111, 1922 [Chapter 6 in this collection]. It has also been established for this case that the formulae for the second part of the light postulate are not true.

Other evidence also supports the special position of the speed of light. At the beginning of this section we defined the geometry of space and time with rigid rods and clocks, but it is not necessary to follow this path. It is possible to completely define the spatio-temporal metric of the special theory of relativity using only optical signals. We imagine an empty region of space or a region that is free of material points; it is then possible to use light signals to define when two points are at rest relative to one another. We can speak of an *optical rigidity* which is unambiguously defined. It is not necessary for this to suppose that you know the uniform time in advance. To the contrary, this is unambiguously defined by the light signals. Furthermore, congruence can also be determined by optical signals. A pure *light geometry* can therefore serve as the basis for the measurement of space and time. The fundamental hypothesis of Einstein's kinematics resides therefore in the statement that light geometry is identical to the geometry of rigid rods and natural clocks.[10]

Herein lies the unique position of the speed of light. This becomes just as clear if a mechanical model is used to represent the kinematics of the theory of special relativity.[11] Such models satisfy the regulation of clocks using a speed slower than light. Since simultaneity can be defined in a completely arbitrary fashion, this is unproblematic. A problem appears when the size of the model is chosen to be small enough that it can be taken in with a single glance. Then the simultaneity of the system becomes identical with psychological simultaneity and the other system appears to be "false." To the contrary, when the size is large, we no longer have psychological simultaneity. If therefore one abstracts from

[10] Compare with the report cited above, "Eine Axiomatik . . ."
[11] Cf. e.g., E. Cohn. *Physialisches über Raum und Zeit,* Teubner, Leipzig.

the smaller size, we obtain a good representation of the larger sizes. One difference with Einstein still remains; in the model, the clocks and rods must be artificially corrected, whereas this is not necessary with light signals: the regulation is automatic (aside from the point of origin for time measure). So once again we clearly recognize the unique role of the speed of light.

Moreover, it should not be surprising that the speed of light is precisely that which plays a unique role in nature. In effect, electromagnetic waves are a phenomenon that has a greater importance than any other signal in nature. They alone allow, ignoring gravitation, the transmission of an action through empty space. The forces that the individual material particles exert on one another are of the electromagnetic variety. All propagation of material interaction is therefore accomplished, in the final analysis, by electric transport: it is therefore appropriate to say that this transport has a special place in nature.

II. The Principle of Relativity

It is well known that forty years earlier Ernst Mach had already expressed an idea essential to the theory of relativity. Mach stated that motion is always defined by a relation between bodies and that it is meaningless to speak of motion with respect to space. In his critique of Newton's principles, he said of motion: "We thereby recognize that to speak properly requires a relation between the body *K* and the bodies *A*, *B*, *C*... If, on the other hand, we ignore *A*, *B*, *C*... and if we want to speak of the state of a body in absolute space, we are twice mistaken. First, we cannot know what comprises the state of *K* in the absence of *A*, *B*, *C*...; further, we have neither the means to judge the state of the body *K*, nor can we confirm our assertion which therefore eliminates any scientific significance."[12] This passage contains an extremely clear criticism of the doctrine of absolute motion and further illuminates the only possible path to a rigorous theory of motion. The idea that motion as a spatial phenomenon can only be understood as a relation between bodies pre-dates Mach. Leibniz, for example, had already expressed a similar notion. Motion, as that which is characterized by a *change in spatial distance*, is relative. This is inherent in the concept of kinematical motion, but Mach's position is strongly distinguished from previous conceptions by the idea that we also have *relativistic dynamics*. Motion can still be acknowledged to be the result of forces – this is the meaning of Newton's equation: Force = Mass·Acceleration – and Mach teaches that the effects of indirect forces, the so-called inertial forces, are the result of the presence of other bodies. Before Mach, this idea was unknown. Mach deduced this from the fact that the entire conception of motion must be reversible and that it must always be possible to interpret inertial forces as the influence of other bodies. Mach clearly expressed this idea for the case of rotation: "But, if we remain in the realm of facts, one can only know *relative* spaces and motions. If we omit the background 'space' which is unknown and makes no sense to us, whereby the motions in the system of the world are relative, they are just like the conceptions of Ptolemy and Copernicus. These two conceptions are equally *true*, but the latter is uniquely simple and more *practical*. The system of the world is not *twice* given (the Earth is moving and the Earth is at rest), but only *once*, since only relative motions are individually accessible. We cannot say what would happen if the world were not rotating. We can only interpret the different manners in which the particular case is given: if we interpret it in

[12] Ernst Mach, *Die Mechanik in ihrer Entwickelung*, Leipzig, 8e Aufl., p. 224.

a way that contradicts experience, *our* interpretation is false. The principles of mechanics allow this sort of account of the source of centrifugal forces via relative rotation."[13] It is amazing to see the certainty with which the idea of relativity is enunciated here. Only someone like Einstein could recognize the genius of Ernst Mach. In his eulogy of Mach, Einstein himself cites these passages of Mach's mechanics.

But is Mach's conclusion correct? Is it true that general relativity is a *logical necessity*? We believe it is possible to reach this conclusion from the fact that we can conceive all observable motion of a system of bodies A in relation to a system B as the inverse of a movement of B with respect to A. One could just as well attribute the observable facts, i.e., the presence of forces, to the motion of A as to the motion of B. Indeed, this is correct, but Mach's conclusion goes too far. Experiments that contradict general relativity can be *imagined*, for example:

Let the universe consist of two separate, but identical, systems S_1 and S_2. We describe the state of motion of the universe in the following fashion: Consider the sphere of fixed stars of system 1, F_1, to be at rest while the Earth of this system, T_1, rotates. The axis of rotation is the line $T_1 T_2$. In the same way, we define the motion in system 2 with respect to a chosen state of rest; it will therefore be possible to judge the motion of the second system from the point of view of the first. In the second system, let it be the sphere of fixed stars, F_2, that rotates with the same angular velocity and in the same direction as T_1 and let T_2 remain immobile. This state of motion will be referred to as "conception 1."

We can also describe the state of motion inversely by supposing F_2 to be immobile, then T_2 rotates while T_1 is at rest and F_1 rotates.

[13] V. loc. cit. p. 226.

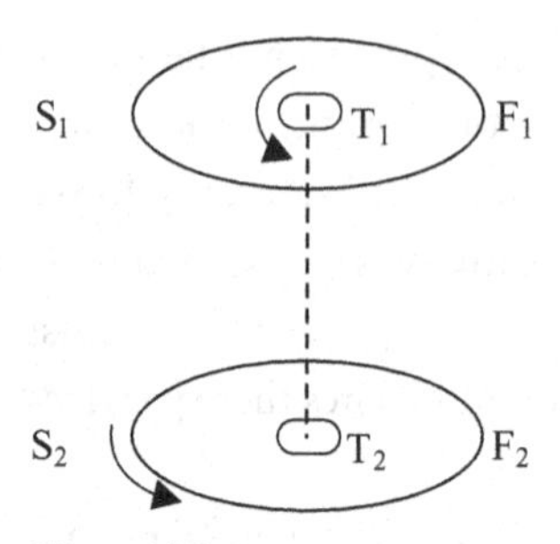

Figure 10.1.

This way of representing things will be termed "conception 2." We see that in every case, F_1 and T_2 on the one hand, and F_2 and T_1 on the other, are at rest relative to one another. From the kinematical point of view, there is no difference between the two. The two are equally justified. On this point Mach is entirely correct, but we must observe the following very remarkable fact: in T_1 centrifugal forces arise, recognizable, for example, by the tendency of some tight springs to bulge; in T_2 these forces are not apparent.[14] There is no doubt that this state of affairs is *possible* (although it contradicts the theories of relativity) because one cannot infer *a priori* that this experiment does or does not allow the determination. How are we to interpret this fact?

In conception 1 we can say: centrifugal forces only arise if the body concerned (T_1) is itself rotating. If, on the other hand, fixed stars are rotating (with T_2), there is *no* centrifugal force. This is the statement of Newtonian absolute rotation.

Can this statement be contradicted if, like Mach, we take the point of view of conception 2? We must again accept that

14 We leave open the question of knowing if the centrifugal forces arise in F_2. We can continue our reasoning without hypothesizing on this point; it will conform to the spirit of our example to admit that there appear equally in F_2 forces corresponding to centrifugal forces.

conception 1 is not logically necessary and interpret the centrifugal forces as an inertial effect of the rotating bodies. From the point of view of conception 2, it can be said: It is only through the rotation of the fixed stars that centrifugal forces arise (therefore in T_1). These forces originate in the rotating sphere of stars. These conceptions of centrifugal force are equally justified.

But we now see a difficulty setting in. From Mach's point of view, we cannot comprehend why a centrifugal force does not appear in T_2, since T_2 is found to be, with respect to F_2, in the same state of relative motion as T_1 with respect to F_1. The operation of centrifugal force cannot therefore be interpreted as the action of relatively rotating bodies. The hypothesis justified in conception 2, that the fixed stars produce an attractive field by rotating, does not conclude the matter because this action *is itself* brought into play only if the stars are found to be in a state of motion absolutely determined *with respect to space.* This is the result of the absence of centrifugal forces in T_2. Although the two systems are entirely alike, they are dynamically distinct.

Against such a conclusion it can be objected that the two systems S_1 and S_2 are not isolated from one another and that in the inversion of the motion we must consider the entire ensemble of bodies. This is correct, but does not get around the difficulty. T_1 and F_2 can still be seen as being combined on the one hand and T_2 and F_1 on the other to form systems S and S'. Two perfectly symmetrical systems in relative rotation are obtained, but in S we observe actions not seen in S'. Relative rotation among bodies is therefore not a sufficient cause for observable phenomena.

The difference can therefore be attributed to a rotation with respect to absolute space. S and S' are found to be in different states of motion with respect to absolute space; this is why the effect of their rotations is different. Absolute space obtains the status of a sort of non-material entity with physical properties (agreeing with the conception of Newton), since it has the property of conveying a physical effect to rotating matter. The only surprising thing is that this effect is a unique characteristic of our space. Since it is possible to hold physical notions other than our own, it is not a logical impossibility.

All the same, it should also be remarked that relativity could be interpreted in another sense because in the world we have just constructed, we will not yet be in a condition to decide whether S or S' is at rest with respect to absolute space. Even in this case, where we can straightforwardly speak of absolute space, we cannot judge the state of motion of matter with respect to this space. Indeed, we are never required to believe that centrifugal force is necessarily an inertial effect, it can also be a gravitational effect produced by the rotation of the fixed stars without reference to absolute space. Newton's philosophical error was to believe that conception 1 alone is possible in the imagined physical conditions above; furthermore, it was Mach's error to have excluded certain real states of the sort we have constructed in the name of relativity.

The difficulty is resolved by drawing a distinction between *epistemological* relativity and *physical* relativity. Epistemology allows for the assertion that every action that is reduced to motion can also be explained as an action in the relativistic conception. But it does not follow that every difference between the *actions of motion* can be reduced to a difference in the *distribution of masses*. It is this second assertion only that excludes absolute space, that is to say, a non-material cause of physical action. By further adding this more restrictive assertion, we move beyond the demands of *epistemology* and accept *physical* relativity. In the problem of motion, as with the problem of time, the solution resides in a distinction between epistemological relativity and physical relativity.

One can now understand why the problem of relativity remains obscure. The two forms that we have pointed out are often conflated. This is also the reason why Einsteinian relativity always appears so particularly strong, so immediately illuminating, that many of its supporters want to derive in an *a priori* necessary fashion what Einstein himself always emphasizes as its experimental origins. The apodictic character of relativity in the epistemological sense is found in these views to be transferred into the content of physical relativity. Einstein posits a theory of *physical* relativity. It is from this alone that it is capable of discovering new observable phenomena. Already, Mach, who was confounding the two sorts of relativity, was led to conclusions about the behavior of rotating disks, i.e., that they produce in their interiors a weak attractive field if the centrifugal forces are created only by the relative rotation of matter. This is clearly the physical sense of relativity. It follows from this result that in each *part* of the universe, independent of its orientation, the same actions appear as are found in the complete system, the difference being only a matter of degree. Also in the case of the example that we constructed, this theory requires the appearance of centrifugal forces in T_2. It is by this sort of reasoning that Einstein deduced from physical relativity that the light of the sun should have its spectral lines shifted towards the red. But it is clear that this was only possible because his theory is relativistic in the narrow sense, that is to say, because it claims that the dynamic effects depend only upon the distribution of masses. On the other hand, the epistemological principle of relativity does not permit the derivation of observable results; it only concerns the different means at our disposal to describe observed empirical phenomena.

Here, we add a remark that would be quite banal if philosophers were not always disagreeing with it. The remark is that the singular characteristic of *uniform* motion, as Newton and Einstein show, is clearly nothing but an experimental result. There is no fundamental physical principle which physics requires for a particular application. That it is possible to have motion without force is purely a lesson that can only be taught by experience. Simple intuition demands that *all motion*, even uniform, find its source in some force. The truth is that we have the habit of only regarding force as the cause of *changes of motion*, but this is certainly only a result of a *process of education* imposed on our understanding by physical experience.

III. Space

The concept of time received a decisive correction in the special theory of relativity. The concept of space, on the other hand, will only be subject to modification in the general theory. On this point, the research of Einstein is connected to the problem of congruence, a problem which, like that above, has been thought to be obvious.

In what sense is the length *AB* the same as that of a length *CD* placed at a distance? There are two responses to this question, one mathematical and the other physical. Here is the mathematical response: the lengths are equal if they satisfy the relations contained in the axioms of congruence. But how do we know if these relations are satisfied? The mathematician easily responds to this question. It must, in fact, be that the lengths in one way or another will themselves be *given*, for example, numerically by the coordinates of the end points, and then it is easy to verify if the relations are satisfied. When the lengths are not given by coordinates, they are the shortest distance between the points or the figures (for example, as lines in a particular triangle), and there then exist logical means for determining if congruence is or is not satisfied. Thus, mathematicians will not see a particular problem with the *verification* of congruence. The *definition* of congruence is no longer a problem for them. It is arbitrary, and what is congruent in one geometry is not necessarily congruent in another.

What makes the physical response more difficult is that the object of the physicist is not given by the content of the relations, but by the perception of facts. A physical length is the distance between *these two points* on *this particular rod*. Take, for example, the measurement of a piece of land. Here we have the question: Is this length the same as the other? Instead of *calculating* as in the mathematical fashion of determination, we introduce *measurement*, that is to say, an operation carried out on physical objects. This leads to a corresponding change in the definition of congruence. This is not a result deduced from abstract properties, but an immediate restriction upon them. It can be expressed in this way: the former length is congruent to the latter length, and just as in the mathematical case this definition is arbitrary, it is irrelevant whether we have designated congruence for this size or another, it is only after such a *convention* is set down that a *measurement* can be made.

It can be objected that for rough measurements the lengths are directly given and intuitive, while for more accurate physical experiments where many corrections are made, length is no longer intuitive but is *defined* by physical laws. It is in this way that the length of a rod for precise geodetic measuring is a number that depends upon the complex behavior of a real rod. However correct this idea is, it does not change the central problem. Because in the precise measurement, just as with the other, the final empirical determination is taken on the basis of sensible intuition, of the perception of a material configuration. A perfect measurement consists only in that we first observe many perceivable things (for example, other standard rods, the meniscus of a column of mercury, the level of a bubble, etc.); and second that the abstract length is calculated from observed sizes such that large variation in the latter lead to only small differences in the abstract length. For example, the variation of the meniscus of a thermometer by several millimeters entails a variation of only tenths of a millimeter in the abstract length of the rod. It is in this profound sense that all of our procedures aim to augment our precision, but because the eventual, more precise measurement itself also reduces to sensible perception, in the course of the following considerations which concern a question of principle, we can separate this process towards precision from its causes and be satisfied with direct measure by intuition because, all the same, it comes back to taking as a standard a directly given sensible length or a complicated function of this length.

We must understand why it is a matter of pure definition when we say that a rod at some place in space is the same length as one someplace else. What are the constraints on this equality? Suppose that there are two rods at the same place and that when superposed they are of the same length. When I take one of them to another point, is it still as long as the second? Is it not possible that during transport a force X acted upon it and altered its length? I transport the other rod to verify this point, but is it not possible that it will also be subject to this force that results in the same modification? There is no way to gain evidence for the existence of this force X if it affects all bodies in the same fashion, but it is for this reason that it may be neglected. It is simpler to neglect it and to call a rod self-congruent at different points. Moreover, I can also say that this rod becomes half its size after displacement, two-thirds its original size after two displacements, etc. This cannot be wrong, I have only introduced here a contracting force X.

We must again add a supplementary condition here. At different points, the rod cannot be exposed, for example, to different temperatures. Physics provides a series of such perturbational influences which all affect the rod. We intend for the rod to be "without modification," a rod which is free from all *known* physical influences. This is, to be sure, only a *provisional* definition that will be further perfected later; it contains a *subjective* element, since our knowledge of natural forces is implied in this definition. It again allows us to freely dispose of the force X because we cannot account for it among known forces. It must be noted that the definition of *solid body* is also made in the same fashion, rigidity is defined in terms of known forces. A deformation due to the influence of a force X at this point remains possible.

Congruence of rigid rods under transport is therefore defined in the presence of an arbitrary field X. The most simple choice is to set $X = 0$. This definition of congruence is arbitrary, but it is univocal, and it entails that two rigid rods that are congruent at a point remain congruent at all points. This is an axiom that we can consider to be experimentally well confirmed.

Once the definition of congruence is given, the following question can be posed: Which geometric relations are compatible with this congruence? Is the geometry obtained Euclidean or not? This question has a completely clear meaning. I use my rods to form six equilateral triangles, and I arrange them at another point in such a fashion that each of them is contiguous with the next. Is the sixth triangle also adjacent to the first? This is a question of fact, one cannot prejudge the response. If experience says no, then the geometry obtained is non-Euclidean.

The following criticism can be made: if the triangles are not connected it is just that the edges have undergone a contraction. The analysis of geometric relations is required therefore in order to decide whether they are subject to a mysterious force X that renders the measurement uncertain. The deviation from the Euclidean relationships allows me to calculate the force X, which is zero here only if the geometry one ends up with is Euclidean.

In fact, a parallel means of argument is *possible*, but not *necessary*. The inverse can also be confirmed, that the normally valid geometry is a Riemannian geometry of completely determined curvature; all deviation from the relations of this geometry can be interpreted as a result of the force X. It is not possible to decide between these alternatives in an absolute fashion. One can only say that if a definition of congruence is made the geometry is determined, and if a geometry is supposed then the congruence relation is found to be determined. In the latter case one can introduce a force X to explain the abnormal behavior of the rods; in the first case this sort of force is avoidable.

Consider an example of these two possibilities. It is possible to determine the geometry of the surface of the Earth with rigid rods. It is not essential here that we are dealing with a two-dimensional measurement, it is meant only to facilitate intuition. Begin by choosing the first procedure: we agree that a given rod always maintains the same length. This results in a two-dimensional geometry that ceases to be Euclidean because the surface of the Earth is a spherical surface. Now use the second procedure: we impose the conditions of Euclidean geometry; this condition can only be satisfied if we introduce a force X which deforms the rods in such a fashion that they give the *appearance* of indicating a spherical form. Since the second procedure in this case appears to be absurd and the force X is artificial, in the three-dimensional case should it not be the first procedure only which is allowed and which becomes the standard? Consider the properties of the force X. It will be easy to experimentally establish that it has the following two properties:

a) The force X acts in the same manner on all bodies regardless of their nature
b) There do not exist insulating walls for the force X,

this is clearly recognized in the fact that in *reality* the geometry ceases to be Euclidean when we select a sphere. We must therefore apply the same corrections to all rods regardless of composition, and we cannot avoid making these corrections despite isolating walls that may enclose a rod. It must again be added that a *known* physical alteration is not to come from the walls (as we have postulated our definition on the absence of such modification) that would compensate for the force X. This restriction is indispensable for the definition of that which we call *isolating* walls. A force that possesses these two properties is said to be *a force of type X.*

It is exactly these two properties which make the force X appear to be fictitious because it entails indeterminacy in all practical measurements. If one allows a force of this sort, one can confirm that the Earth is a cube or any other arbitrary form. There is no longer any sense in speaking of the geometric form of a body when one allows for a field due to forces of type X. It seems therefore to indicate an acknowledgment in a general manner that forces of type X are not admissible. This is the convention that we implicitly make when we say that the Earth is a sphere. The case of three dimensions is to be treated no differently than the case of two dimensions. Here we also make the same convention to avoid the reproach of fictitious forces.

In this way, we finally arrive at a complete determination of the concept *without modification*, the definition of which has the advantage of removing all subjective elements. A body can be said to be transported without modification if:

1) It is not subjected to known physical influences;
2) Forces of type X are not admitted in the comparison of lengths at different points.

Thanks to this addition, the indeterminacy that results from our imperfect knowledge of natural forces is reduced to a small remnant of inexactitude that is inherent, in principle, in all physical measurements and in all definitions related to measurement. On the other hand, a force of type X is, in general, more influential than this remnant. For example, the force which makes the practical measurements on the surface of the Earth turn out to be Euclidean in two dimensions can be calculated with a degree of precision, in general, more accurate than experimental error.

The role of geometry in physics is yet to be elucidated in a complete manner. Geometry is arbitrary as far as no convention has been made concerning the concept "without modification," but it is possible to clarify this concept in a purely physical fashion in relation to the properties of forces and without employing geometry. Rigid bodies can be defined in a natural fashion without it being necessary to impose some geometry as the normal geometry. Once this is done, the question of geometrical determination becomes experimental. But if this is not done, geometry, and the shape of physical objects, remains indeterminate.

Note that only now do certain assertions of the special theory of relativity become meaningful. Einstein has often been opposed with the claim that clocks which are slowed cannot be invariable, giving rise to the criticism that the moving system is served by other units of measure than those of the stationary system. But it is necessary for Einstein that the units of time and the units of length of the moving system are defined in the way that is known to avoid the coming into play of forces of type X. By *a second moving system* is meant the unit of time of a clock that is transported from the system at rest to the moving system. This results in the retardation of the moving clock. If to avoid this slowing, the clock is corrected by dividing its rate by

$$\sqrt{1 - \frac{v^2}{c^2}},$$

the deviation of the clock with respect to the normal rate must be interpreted as the result of a real force that is, as one can clearly see, of type X. Einstein's units are therefore again constrained by the theory of special relativity chosen to be of the sorts that do not give rise to forces of the type X. We call such units *natural units.* It is only now that we have the exact definitions of invariable units and isolated systems that were alluded to above (section II). These definitions result in the exclusion of all *known* influences and all forces of type X.

The fundamental importance of this way of thinking is found in the theory of gravitation, because gravitation, which until now has been thought to be like every other force, is precisely, considering the discussion above, a force of type X. It is known that it affects bodies in the same manner (the equality of inertial and gravitational mass) and that it is impossible to isolate against the effects of gravitation. The thoughts of Einstein have therefore brought us precisely to the consequence that gravitation *as a force* may be eliminated, and in its place we put *the spatial metric* (more precisely that of four-dimensional space-time). With Einstein we no longer say: a planet describes a trajectory which itself is assigned because it is attracted by the force of the sun, but we say: the planet moves along the straightest line that is available in curved space.[15] It is also in this fashion that we can obtain the remarkable and controversial identification of gravitation with certain fictitious forces that result from a change in coordinates. Imagine a gravitation-free space in which we find a system of coordinates formed by a network of congruent rigid rods. In this space, the metric is Euclidean, that is to say the $g_{\mu\nu}$ possess the well-known special form. If we now introduce new coordinates such that the rods get progressively shorter the farther to the exterior a rod is, so that the network will be considered to be curved, then the new $g_{\mu\nu}$ take a form different from the special form. This can be conceived of in the following way: there exists a force which shrinks the rods and this force is represented by the deviation of the $g_{\mu\nu}$ from the special form, it is therefore considered to be a correction in the establishment of the ds^2. It is clear that this force is of the type X because it is only a *fictitious force* produced by the anomaly of the rods. All of the magnitudes that, in an element of the network, are measured with a local unit undergo the same correction; this is precisely the motivation for considering this fictitious force to be interpreted as a gravitational field. The only generalization that

[15] Note that one must allude here to a continuous space-time of four dimensions. The projection of a geodesic of four-dimensional space, i.e., a Keplerian ellipse, into a three-dimensional space is not a geodesic of the three-dimensional space. Further, it is necessary here to draw a distinction between the gravitational potential and the gravitational field. The gravitational potential (the $g_{\mu\nu}$) has only an immediate metrical significance. Its gradient, the field, can change with the construction of clocks. It is in this way that the sweep of a pendulum clock at the equator does not run at the same rate as a spring-based clock. To the contrary, Einstein's red shift that is an effect of this potential is the same for all clocks.

is presented here is that the gravitational field is represented as an elastic tensor field not a scalar potential field. But this generalization again is rendered necessary by the introduction of inertial forces in the concept of gravitation. It is quite remarkable that both the gravitational field and the corrections resulting from a simple change in coordinates can be brought together in a single concept. This holds both for the latter which, as metric corrections, always possess the properties of the force X and to the other part, gravitation, which is ordinarily recognized as a physical force, yet possesses these properties. This is of profound significance, it says that gravitation reduces to the metric.

But we must again insist on this point: the definition that is given for rigid bodies is not *necessary*. One can just as well admit forces of type X and prescribe an arbitrary geometry. It is a matter of fact to be determined given the value of the field X. This process is only an inconvenience, i.e., the geometric form of bodies ceases to be objective. But even now a physics can be constructed which is objective for the structure of the field X. We observe that the real problem lies in deciding between these *alternatives*: *either* Euclidean geometry *and* a field X *or* the geometry determined by experience and *no* field X. This sort of response is characteristic of the epistemological solutions for which we are indebted to the theory of relativity.

The solution to the problem of space is therefore found only in this conception we call conventionalism and which goes back to Helmholtz and Poincaré. But one has no right to abuse this conception in order to diminish the importance of Einstein's discovery. One often reads that Einstein simply showed that it is *possible* to conceive of space as non-Euclidean, that this renders physics particularly simple. The first assertion is not correct because the *possibility* of conceiving of space as non-Euclidean has already been shown by the two physicists mentioned above, and the second assertion is false because the simplicity obtained by retaining Euclidean geometry is not very deep. To use a field of force X does not complicate physics in any essential manner. This is the most profound result achieved by Einstein. When he asserted that space is curved, this is true in the same sense in which we say that the Earth is a sphere. If one wants to avoid the geometry of Einstein in order to escape relativity, the spherical shape of the Earth and the general manner of all geometric shapes of real objects must be renounced.

This is the path that Weyl followed with perfect rigor. The interest that motivates this extension of the theory of relativity is that one can recover the duality between physics and epistemology. From the point of view of epistemology, one can show that all definitions of congruence make possible a certain physics. It cannot be said that a rod is equal in an absolute sense to one at another place, but that we can dispose altogether with the arbitrary process of the comparison of rods. This is the sense, from the *epistemological* point of view, of the claim of the *relativity of magnitude.* But Weyl goes beyond this claim. He requires that *for every definition of the comparison of magnitudes the laws of physics maintain the same form*; this is the *physical* claim of the *relativity of magnitude.* It signifies that the physical systems which are possible from the epistemological point of view and which are given rise to by the various different definitions of the comparison of magnitudes are equivalent. However ingenious is the development of Weyl's mathematical theory, its exactitude can only be confirmed by experience. The parallelism with Einsteinian relativity of the systems of reference, as indicated by Weyl himself, is imposed on the mind. This is precisely the motivation behind Weyl's theory that requires experimental confirmation. Unfortunately it is impossible at the moment, according to Weyl, to decide this point.

IV. The Problem of Intuitive Evidence

It is necessary to be perfectly rigorous when formulating concepts. This is why it is scarcely possible to further object to the relativistic concepts as they have been developed. But the opponents of the theory of relativity, while agreeing to these concepts, assert that Euclidean geometry still holds a privileged place from the point of view of intuition. This is the point that they attempt to clarify. To tell the truth, one thing is certainly correct. All the same, if Euclidean geometry does hold this privileged position with respect to the evidence, then the physics that is imposed by a Euclidean geometry and the field of force X *is no truer* than the physics with a non-Euclidean geometry and no field X. It is by no means a criticism of reality that its character can be precisely described in two ways. It is proper for reality to be invariant with respect to measurement. The Euclidean physics may always possess a *psychological* advantage, independent of the character of truth, as the one that is the most intuitively evident, but this is precisely what we need to see.

As long as one is limited to the intuitive character of real things, a response is easy to give. I imagine eight cubes which, when measured with a rigid rod, have all sides of equal length and precisely right angles. I can surely represent to myself intuitively that in the superposition of these cubes (all being of a common height) their surfaces do not coincide. As it is a question of real bodies, I can have no idea as to how they in fact stack up upon one another.

Through an "empirical intuition" (in the sense of Kant) it is possible to represent non-Euclidean metrical relations.

It may be objected that I have deformed the cubes in transporting them. Certainly it can be expressed in this way, it is physics *with* the field X. But why am I obliged to call this influence from this field X a deformation; could I not make use of a physics *without* the field X but give up the effect of a force with the strange properties of a field X and speak of a non-Euclidean geometry? Logical necessity does not require it. Euclidean geometry is not a *condition of knowledge* in the sense of Kant because we are also able to construct a system of knowledge with a non-Euclidean physics without a field X.[16] This may seem contradictory, it is counter-intuitive. I cannot *imagine* that invariable cubes will not coincide. It is therefore intuition that requires a deformation, but not of course the *empirical* intuition: this only affirms the simple *fact* that the cubes do not coincide. We agree with Kant in recognizing as the source of our tendency a *pure intuition*. Everything leads back to the problem of this pure intuition. Whether or not the rigid cubes satisfy Euclidean or non-Euclidean metrical relations (in fact, it may well also have Euclidean relations with the field $X = 0$, but this is a point which we can never know *a priori*), both can be equally well represented from the intuitive point of view. It only seems intrinsically necessary to speak of a deformation of the cubes in one case but not in the other. We call this necessity *pure intuition*. In what does it consist?

It happens that mathematicians who are well trained in non-Euclidean geometry say that this geometry becomes increasingly

[16] From the *pedagogical* point of view, it is no doubt better to pass over the school of the Euclidean physics, we do not contest this; but this is different from Kant's problem.

intuitive to them, and those who work with the theory of relativity also move away from the ordinary argument, saying that the *intuitive* character of Euclidean geometry is only a product of habit and that non-Euclidean geometry can, little by little, become intuitive as well. Here we bring ourselves to yet another point of view. I would like to show that it is an error to believe that of the two geometries, the Euclidean alone is intuitively representable while the non-Euclidean is *essentially counter-intuitive.* In order to show this, I speak only of geometric forms as objects immediately given through perception, since, as we have seen, *empirical* intuition is not the origin of our urge for Euclidean geometry. It is only a question of pure intuition. To pursue this question, consider the following example.

I want to represent a sphere the size of the terrestrial sphere; how can I do this? I envision a small sphere, for example the size of a globe; on this moves a person the size, say, of a toy soldier. Incontestably this is not the ratio that actually obtains. I seek to enlarge the sphere, to make it as large as a room, as large as a mountain, but never as large as the Earth. Why can I not represent these dimensions? Let us try another. I conjure an image of the sea; from the horizon emerges the smokestack of a ship. In this way I represent the curvature of the Earth. Of course, this appearance is exaggerated because I underestimate the distance of the ship, but I can always correct this inaccuracy. I stretch then, in every sense, this curved surface as far as I have said. What is the result? A sphere whose radius is more or less the depth of a mineshaft.

Please excuse these psychological thought-experiments. They are only used to illustrate this simple fact: a sphere the size of the Earth is not intuitively representable. One can envision through intuition neither the extremely small curvature of an element of the surface of this sphere having a size on the order of a table – if I represent this curvature to myself, the curvature is too large – nor the enormous surface area of the entire sphere. What does it mean to say therefore that a sensible sphere can be represented intuitively? It means that:

1. I can intuitively represent a *small sphere*;
2. I hold that the *large* sphere is geometrically similar to the small sphere.

I do not object to the first proposition. But where does the second come from? It can be supported in two ways; either it is logically necessary or it stems from the needs of the intuition which we have been analyzing.

But it cannot be logically necessary because of the long-established relative consistency of non-Euclidean geometries which deny proposition 2.[17] Further, this assertion is not an empirical condition as was stated above. All that remains is the second alternative: the requirements of an intuition.

But from this point of view, pure intuition leads to serious difficulties. Because proposition 2 sets out a relation between the large sphere and the small, and as the large sphere is *not* intuitively representable, intuition cannot teach us much about the *comparison* of the two. One is reduced to looking for some sort of different way out to satisfy the needs of the pure intuition. One could say: For any intuitively given sphere, I can correspondingly represent to myself a larger sphere[18] and for these two spheres intuition is obliged to admit a geometric similarity. It must also follow that the very large sphere is geometrically similar to the small. But this

[17] In non-Euclidean geometry, there do not exist similar figures of different sizes.

[18] Remark: One *cannot* represent a sphere as large as one would like because intuitively representable spheres have an upper limit which itself is not accessible.

"must follow" is a conclusion by analogy which does *not* emanate from intuition. It maintains that there is an analogy between the intuition of "the small" and of larger dimensions, an assertion that is not itself found in the intuition of small and which remains an extrapolation.

Here now is the decisive point. Proposition 2 does not hold absolutely for the relations of the *intuition*. It sets out a relation between the rational configuration "large sphere" and the intuitive configuration "small sphere." But this relation is not the result of an *intuitive* process because the comparison of the two spheres does not come from the *intuition*. Proposition 2, therefore, cannot be required by the intuition because it is not itself realized in the intuition alone.

But if proposition 2 is neither logically necessary nor comes from the intuition, where does it come from? It is, no doubt, a matter of simple *habituation*. Yet, we have not followed Hume in saying that this habit is mysterious, nor do we fear that we have found within it a contradiction. The only thing that we have found is that in the physics we have developed, the field X is not canceled out. We no longer give proposition 2 a privileged status, nor do we accept it as a synthetic *a priori* judgment. It can always be abandoned and replaced by another habit. It is by no means a statement of fact, but a *definition*, and with this title it becomes arbitrary. We replace it therefore with the following definition:

3. The large sphere is of such a nature that by an intuitively representable reduction of its dimensions, it continuously passes into an intuitively representable sphere.

Manifestly, propositions 1 and 3 allow the construction of a physics that is intuitively representable in exactly the same way as Euclidean physics, because that which is intuitively representable is exactly that which is considered as such in any physics. That which is not representable in Euclidean physics is *not representable* in any physics. A difficulty is again raised: a non-Euclidean Riemannian space, taken through infinitesimally small elements, is approximately Euclidean, but not rigorously.[19] But it seems that our intuition in these domains prescribes a *rigorously* Euclidean geometry. This makes the representation of the infinitesimal elements of Riemannian space not impossible, or inversely, a space in which all of its elements are strictly Euclidean must also be strictly Euclidean in the whole. Thus we cannot therefore find a way to continuously apply the intuition from infinitesimal space to arrive at a non-Euclidean geometry in the large.

This will be exact if the pure intuition of the infinitesimal elements *rigorously* require Euclidean geometry. But as we will show this is exactly what we cannot do. Consider as an example the problem of parallels.

Let us suppose that a straight line is characterized in the following way: If I take a finite segment of this line and displace it in a given direction so that it continues to cover part of its original position, the second segment also lies on the line. This is an intuitive way of characterizing the line that is possible because the movement of geometric figures is intuitively representable. It is not a *definition* because it is not univocal (the arcs of a circle have the same property), but it is a characteristic of the line. During the displacement, the line segment is not deformed, but there are not difficulties here because, unlike in the case of the cubes, we have not made a choice of physics, but merely a mathematical configuration. We understand by "displacement without modification"

[19] Anywhere the curvature tensor is not zero.

just that which the intuition means by this name. *If it is not* the same determination as that of intuition, then the problem will be entirely resolved in favor of the relativistic point of view. We begin therefore by admitting for our demonstration that such a determination exists.

The axiom of parallels says: through a given point on a plane there is one and only one line which passes through it and does not intersect a given line.[20] But this form of the axiom is inappropriate for our context. It is a statement which concerns the domain of ordinary size (it overlooks the important case of lines which do not intersect in the interior of a small region) and we have shown above that for very large domains, intuition fails. We must therefore restrain the domain of this axiom to small regions.[21]

We can define parallelism in the following manner. For an intuitively given line, drop perpendiculars of the same length. Intuition suffices to define the phrase "same length" and "perpendicular." The set of extremities of all these segments are the parallel. The problem which is posed is whether we know if the line obtained in this way possesses all of the above-mentioned properties of the straight line: viz., if the finite segments can be displaced in the proper direction. This result is not guaranteed by definition, it is a synthetic statement. Intuition says that it is so, but does it have the right to do so?

It can no longer be understood except *approximately* and not *completely rigorously* because the displaced segment can only be

[20] We are abstracting here from the fact that the other axioms (Cf. Hilbert, *Grundlagen d. Geometrie*, p. 20, Leipzig) permit us to conclude that there must be *at least* one line of this type. This proposition no doubt possesses intuitive evidence in the ordinary sense.

[21] It suffices here to limit the axiom to one of its parts, namely the first, because we have to try to find only one case for intuitive evidence that supports it.

represented occupying *each of* its positions. If we try to represent it by a sliding motion, we could nevertheless only see the isolated positions of the segment. The entire set of perpendicular segments could not be represented. This fact is concluded by analogy. We are given only a certain number of positions from which to draw a conclusion about *all* positions, and this conclusion itself is not intuitive.

The following example also shows how much intuition, even pure intuition, is inexact when it comes to a question of judging parallelism. Imagine two parallels as ideal figures. Can we say that they are really parallel? One could respond in this way: "I can again imagine that they are always separated by the same distance." This is something which you cannot *see*, it can only be rationally postulated. For *pure* intuition (pure in the sense that it is not mixed with the rational), there is only one rigorous criterion for parallelism. Two lines that intersect at a cosmic distance cannot be distinguished from two parallel lines in the interior of a region accessible to intuition. This is true, not only for empirically real lines (light rays, for example), but also for lines that we represent ideally. What prevents the designation of these lines as parallel is that we know (not by intuition, as it cannot extend that far) that the lines intersect at a great distance. The parts that we see with the intuition have the characteristics of being parallel. Pure intuition therefore is not rigorous, but approaches it; we achieve rigor only through the application of concepts, at which point the intuition ceases to be pure.

This is why the small regions of non-Euclidean space are intuitively representable in the same sense as Euclidean regions. That which is representable – that for which we know how to conjure an approximate image – is equally representable in both cases, and that which is not representable – rigorously – is only introduced in the two cases through concepts. The mentioned propositions 1 and 3 are reconcilable with them.

Kantians may object that my conception of "pure intuition" as a sort of psychological faculty of representation differs from Kant's usage. I find that every attempt to effectively realize a pure intuition in an *intuitive* manner ultimately returns to this faculty of representation. The conceptual element is clearly separated from the elements of the intuition that the preceding examples clearly show. "Intuitions without concepts are blind"; it is not then that a certain combination of intuition and concepts form "pure intuition." The two certainly pertain to *knowledge*, but in the combination of intuition and concept there are many possibilities that Kant did not believe in. The non-Euclidean concepts can just as well be associated with the intuition as can the Euclidean concepts because the Euclidean *intuition* is also a combination of intuition and concepts. The point of view that there exists no pure intuition was for the first time clearly put forward by Schlick.[22] His approach can be seen to be noticeably in agreement with what has been developed here, it is necessary only to draw one more distinction between intuitive perception of empirical reality and ideal intuition. It is this ideal intuition which I have called "pure intuition," but if one means by pure intuition a combination of concept and intuition it seems that the point of view of Schlick must be formulated in a slightly different way: if there is *one* pure intuition, there are *many*. Among these, the Euclidean intuition is not privileged. It too is only a combination of intuition and concept.

Given that which has preceded, any physics is intuitively representable when its geometry is intuitive in a region accessible to our faculty of intuition. It is once again clear that in very

[22] M. Schlick, *Allgemeine Erkenntnislehre*, J. Springer, Berlin, 1918.

small regions, for example, of atomic dimensions, we must admit large deviations from the relations of Euclidean geometry. These dimensions cease, in effect, to be intuitively representable. It is not necessary that Euclidean geometry always have a differential element, it can appear as an integral phenomenon. In this way a distinguished place is maintained for medium-sized domains defined by the dimensions of the human body where Euclidean geometry is extremely valuable as an approximation. This corresponds equally well to the actual state of our physical conceptions. But it must be noted that this state of things is not necessarily a definitive result and that there may come a day when a modification of our intuition for this domain is required. To apply the term ideal to this intuition must not be understood as meaning that it is a rigid schema. It is possible that this ideal intuition also – limited to medium-sized dimensions and an imprecise metric – contains many rational elements that we have overlooked. It is manifest that we possess the faculty to eliminate little by little the rational elements of the complex intuition; the evolution in modern mathematics furnishes a proof of this. Once the rational elements are discovered, we see that we are free to modify them and give rise to a new combination of intuition and concepts. Ideal intuition is only a word to designate a limit that this process of elimination approaches without ever attaining.

V. The Problem of the *A Priori*

The classical concept of the *a priori* originates with Kant. The philosophical critique of a theory that penetrates deep into the *a priori* principles of physical science cannot therefore avoid a confrontation with Kant. A response of great impact must be made to this confrontation.

It is of decisive importance to note here that in Kant's concept of the *a priori* there are two significantly different aspects, two completely logically distinct components, two interpretations that are usually considered equivalent, but the identity of which represents the essence of Kant's problem. One part of the term "*a priori*" concerns the necessity and universality of certain propositions, and it designates these two properties as the criteria of *aprioricity*. But the other part Kant has set out is that the *object of knowledge*, the physical object, is only defined after the introduction of certain principles or categories which bring together the multiplicity of perceptions and unite them into the thing. This *constitutive* character of certain principles is also for Kant that which provides the meaning of *a priori.* It is a question therefore of two completely logically different points of view: the one concerns the *validity* of the judgments and the other their *place* in the theory of knowledge. Can these two meanings of *a priori* be merged?

Kant answers this question in the affirmative. Indeed, he sees in this response the essential result of transcendental philosophy. Indeed, the point of departure of this philosophy is the question: How are synthetic *a priori* judgments possible? In this question the term "*a priori*" takes its first sense, that of apodictic validity. Kant responds by trying to show that these judgments are the *conditions of experience*, that they alone constitute the physical object, viz., the thing as intelligible. It asserts therefore that the union of the two senses of *a priori* alone permits the solution of the great philosophical problem that has been posed through many years by philosophy, the character of apodictically certain judgments. Furthermore, it was Kant's efforts to render intelligible the long-considered *first sense* of *a priori* that gave rise to the *second* sense.

Now, it is a frequently observed phenomenon that the path followed in the discovery of new relations turns out to be false, despite these relations manifesting themselves in a profound sense. This is also the fate that has befallen the philosophy of Kant. On this point, the decision was a result of the science which Kant himself used as a model for all of science: mathematical physics. We have shown in fact that the first sense of *a priori* cannot be maintained in the face of the theory of relativity, while the second is retained with even more depth and solidity.

The conception of space and time that was developed in the preceding sections are in opposition to those of Kant because they undermine the apodictic character of the structure of these forms of intuition. Indeed, it must be granted that the theory of relativity throws no doubt upon the *internal coherence* of Euclidean geometry. But while Kant calls the axioms of this geometry synthetic *a priori* judgments, he wants to assert more than internal consistency. He holds that these axioms must always be presupposed for empirical space, that which contains the real things. He believes that scientific knowledge is only possible if the forms of pure intuition are employed. It is rightly said here that the theory of relativity has shown the contrary, it has built a scientific system *that does not employ* these forms. It shows that it is possible to have knowledge of nature through *other* conditions of experience than the Kantian conditions. If, all the same, one still wanted to defend the Kantian doctrine of space and time, it would have to be demonstrated that in order to make use of non-Euclidean geometry and relativistic time, one must also presuppose their validity as forms of Kantian intuition. But such a demonstration is entirely impossible. One cannot object that it is possible to represent relative simultaneity in a perfectly intuitive fashion without making use of absolute simultaneity. If, despite this, one still maintains an absolute simultaneity that we have dismissed as inaccessible, it will be difficult to contradict him. It is simply impossible to believe that this absolute time is a condition of experience. Kant himself had certainly rejected a similar "phantom" because to him the form of intuition has sense only because it is not a simple product of the imagination, but is a condition of knowledge. It is the same with Euclidean space. It cannot be said that it is a presupposition for understanding non-Euclidean space. One could only attempt this sort of demonstration by showing that Riemannian space is comprised of Euclidean differential elements. But the Euclidean space of Kant arises as essentially valid only for finite dimensions, and this property by itself *is not* a necessary condition for Riemannian space. It can therefore be said that the theory of relativity has built a scientific theory without employing the Kantian forms of intuition. The privileged place which Kant gave to Euclidean space and absolute time therefore has no *raison d'être.*

It cannot be objected further that by forms of intuition Kant intended more general structures and did not limit himself to precise special forms. It goes without saying for Kant that his pure intuition is identical with Euclidean space and absolute time. Even though he separated the determination of empirical time from pure intuition, making it an empirical problem, he does not want to say that physics can allow all possible definitions of uniformity and simultaneity. He means to say that it is an empirical problem to determine the real mechanism which best corresponds to uniform absolute time and to indicate the empirical procedure that most exactly measures absolute simultaneity. He did not think for an instant that the entire course of time, whatever it is, can be indifferently *defined* as the measure of uniformity. Perhaps he realized that *kinematics* does not constrain us to a single selection, but he was too attached to the terrain of the theory of Newton and Euler not to recognize a means of approximately determining absolute time in a *dynamic sense* in the well-founded selection of the principle of inertia. No doubt one cannnot interpret Kant in this very naïve way and say that he regarded absolute uniform time as psychological time. It is in the well-founded selection of physical laws, that is to say, founded upon the system of knowledge, that he sees as the means of approximately determining absolute time. Nevertheless the fact is that the complete set of natural laws *does not provide* such a selection, they maintain their invariant form under arbitrary changes in the temporal metric. It is a point that Kant did not know because without this fact the theory of relativity would not need to await the coming of Einstein. It is the same with space that is given by pure intuition. Kant intended Euclidean space. The point of departure of his critique is the epistemological question of the source of the apodictic certainty that accompanies the axioms of Euclidean geometry. The answer is that through the pure intuition we are obliged to recognize the axioms. But what sense is there in this response if by the same pure intuition we are obliged to also recognize the opposite, that is to say, the contraries of these axioms? This renders impossible all predetermined choices of axioms, and all synthetic *a priori* judgments about space become impossible. The pure intuition of Kant *is not* reconcilable with the doctrine of space and time in the theory of relativity, and if many neo-Kantians try to conceal this opposition by their choice of appropriate quotations, we speak once and for all in a way that better serves Kant if we abandon the content of these assertions in the face of new physics and if, keeping in line with the spirit of his overall plan, we seek the conditions of experience in a new way. This best applies to those who cling to piecemeal assertions. Kantian philology is no longer defendable today with the new physics knocking on the door of philosophy.

The particular significance of the theory of relativity consists in its shaking not only certain propositions of classical *physics*, but also other propositions which have had a particular nobility for philosophers and are held by them to be eternally immutable. How is this possible? Did not Kant himself give a special proof that *a priori* judgments cannot be refuted? Indeed, he believed so. Since these apodictic propositions (that is to say *a priori* in the *first* sense) have served to construct the experiences (this is the second sense of *a priori*), they must also be employed to construct the experiences contradicting them. Therefore, it is concluded that any such attempted refutation of the *a priori* propositions will start a vicious circle. We want to point out that the knot of Kant's reasoning rests on the union of the two meanings of the term *a priori*.

But this demonstration is false. We can easily show it in the presentation of his forms in a logical schema. Consider the hypothesis: *A* is *B*. I demonstrate by starting with it and by employing other auxiliary propositions: *A* is not *B*. This is really a complete proof of the inexactitude of the proposition: *A* is *B*, as it leads to a contradiction. This is a means of demonstration frequently used in mathematics. It is therefore allowable to permit, in the premises of the denial, the statement of the denial itself. This shows that the inexactitude of the constitutive principles can sometimes be demonstrated. One can, for example, demonstrate in this way the invalidity of Euclidean geometry. Indeed, it should not be understood in the sense in which a representation of reality in terms of Euclidean geometry is *impossible*. Again we have shown above that to completely maintain the amenability of certain *supplementary* conditions, it is necessary to use non-Euclidean geometry. But if one holds onto the principle that congruence is to be defined by solid bodies, the invalidity of Euclidean geometry can be shown to exactly follow from the schema above.[23] The fact that Euclidean geometry remains *possible* cannot be regarded as saving the Kantian *a priori*. What is decisive here is that the validity of a *combination* of several principles can be a matter of empirical inspection. Kant's critique gives us an entire system of *a priori* principles, and it cannot be determined in advance whether this system will eventually find itself irreconcilable with experience. Furthermore, this result is in complete agreement with the view developed above because a principle of congruence as defined by rigid bodies, including that of Euclidean geometry, can be considered to be a principle of reason. However, Kant only mentions the one way of expressing it, and when combined with the other rational principles we find it contradicted by experience. It can therefore occur that experience forces reason to renounce the system that gave rise to it. In this way, Kant's demonstration of the apodictic character of the *a priori* must be rejected. We can agree with Kant that certain principles have a constitutive character without regarding them as apodictic. This *union* of the two senses of *a priori* is not at all *necessary*, and the second notion, the proper discovery of Kant, does not have the firm foundation of the first. Their *separation* remains the sole means of protecting the concept *a priori* from the progressive attacks of experimental science. This immediately gives rise to a rather difficult problem: How is it possible, after certain principles have been recognized as false, to arrive at new principles? This problem cannot be solved with the help of the logical schema sketched above because although I can derive the proposition "*A* is *C*," I cannot include the statement "*A* is *B*" into the premise set of the *direct* proof because it effectively results in a contradiction.

23 I have shown this in complete detail, *Phys. Zeitsch.*, vol. XXII, p. 379, 1921 [Chapter 3 in this collection].

Theoretical physics has arrived at a logical process that allows us to elude this new difficulty. It may happen that in certain simple cases we can admit the *approximate validity* of the proposition "*A* is *B*," and then show that there is another law, "*A* is *C*," that is in fact true and which approximately reduces to the first law, "*A* is *B*" in the simple cases. In other words, there is a place for *continuous expansion* of the constitutive principles. This logical process is called the *method of continuous expansion.* It should be pointed out that this is the path followed by the theory of relativity. When Eötvös experimentally established the equality of inertial and gravitational mass, he had to, in order to calculate his observations, admit the validity of Euclidean geometry in regions on the order of his balance. In spite of this, the results of his inductions were able to provide a proof of the validity of Riemannian geometry for a system of cosmic dimensions. The corrections to the measures of length and time required by the theory of relativity are of such an order that they can usually be neglected under normal experimental conditions. When, for example, an astronomer moves the watch with which he takes his measurements of the stars from one table to another, he does not need to introduce *Einsteinian* corrections for the watch. Nevertheless, he can use his watch to reveal a position of Mercury, which implies a displacement of the perihelion, and thereby demonstrate the theory of relativity. When the theory of relativity asserts that light rays curve in the gravitational field of the sun, this does not prevent either the observation of the stars under the supposition of a straight optical path within the telescope or the calculation of the corrections to the aberration by the usual method. This process works not only when we draw conclusions about the larger from smaller dimensions. If, through the progression of the theory, we arrive at the prediction that for an electron there is a large curvature of the space in the interior of the force field of the electron, this curvature can be indirectly verified with an apparatus whose dimensions are of ordinary scale, and, consequently, whose working principles may be considered to be Euclidean.

In my opinion, this process of continuous expansion is the web of reasoning which stands in opposition to the Kantian doctrine of the *a priori.* It not only provides a way of refuting the old principles, but also a means of establishing the new ones. This is why this procedure seems to properly dispel not only all of the theoretical concerns, but also all of the practical objections.

Abandoning the first sense of *a priori,* the apodictic character, does not force us to also abandon the second sense. To the contrary, the study of modern physics shows that it always makes use of constitutive principles. We must simply get used to the idea that these constitutive principles can be well grounded without the pretension of declaring them to be eternally valid. *Length* has nothing to do with the *validity* with respect to the theory of knowledge. We can only assert one thing, that in the present state of knowledge we use such and such principles for the definition of the object, and we may essentially admit not only that the results of every science but also the concepts of the physical things as real must be subject to a continuous evolution.

Moreover, the modification of the concept of *a priori* is connected to a change in the *method* of epistemology. It no longer makes sense to try to deduce the constitutive categories from reason. Instead of this procedure we introduce the *method of scientific analysis.* The results of positive science, found in constant relation with experience, presuppose the principles whose discovery by logical analysis is one of the problems of philosophy. A large step has already been taken in this direction thanks to axiomatic construction, which, since the geometric axioms of Hilbert, has provided the means for using the modern concepts of mathematical logic. It must be realized that epistemology has no other course than to *uncover the actual principles used in knowledge.* Kant's attempt to deduce these principles from reason should be considered a failure; in place of the deductive method we put an inductive method. It is inductive insofar as it exclusively maintains all of the acquired material of positive knowledge. But this method of analysis does not have a natural relation with the logic of induction. It is to avoid this misunderstanding that we have chosen the name, *the method of scientific analysis.*

In that way the problem of the *a priori* gives rise to a form of solution that could not have been foreseen at the start. In the place of apodictic status we introduced the epistemological point of view, and in the place of the critique of reason we introduced the analysis of positive science. It can be well said that this solution fits well with the tendencies of our times that seek to replace all pretensions of absolutism with logical conditions and rational speculation with work of specialized detail.

Conclusion

The philosophical analysis of the theory of relativity led to a remarkable result. We have been able to establish that this theory, which has helped theoretical physics account for so many experimental facts, leads to a *philosophical* discovery of profound importance. It is necessary to say so much more than the ever-recurring assertion that the theory of relativity is purely *physical,* or as certain extreme opponents of the theory put it, purely *mathematical.* They object that, by definition, philosophy cannot be influenced by the experimental sciences; they do not hesitate to defend this thesis at any price with the most incredible objections to the theory of relativity.

In fact, this thesis is not as faulty as it appears, but the partisans do not suspect that theirs is the least fruitful line of defense. Admittedly, philosophy is sheltered from attacks based upon experience *on the condition that they limit their analysis to the logical structure of scientific systems.* But then it rightly ceases to be in a position to produce any statement about *reality*, about the world of *real things.* Even further, it can no longer posit the very general precepts like those taught by Kant and all aprioristic philosophies. The assertion of a predetermined structure of space and time, of causality, etc., to be as *general* as we wish to consider them, does not constitute reality despite all their *special precepts* simply because more general precepts than these can be imagined which serve as their *foundational concepts. Maximally general concepts do not exist.* This is why philosophy cannot consist of the discovery of a more general structure for reality. Philosophy must leave the edification of these structures to the particular sciences and limit itself to *explicitly* describing the structures that the particular sciences *implicitly* employ. When understood in this way, it must be pointed out that it is by no means the goal of particular sciences to give the structure in question its most general form; to the contrary, science is preoccupied with the choice of general principles that are *as specific as possible.* It is in this sense that it can be asserted, for example, that causality is a general principle of the knowledge of nature. It is always possible to apply this principle.[24] But it cannot in the least be guaranteed that physics will not one day renounce causality.

[24] This principle is by no means *sufficient* everywhere; science proceeds via statistics through a second type of general laws. Cf. *Naturwissenschaften*, p.146, 1920.

If the structure of our knowledge is a representation of reality (in the mathematical sense of correspondence), it is because it is continually adapted to it. For this very reason, the *validity* of that which has been deemed empirical cannot be asserted, and philosophy has no other course but to explicitly formulate its content.

The method of scientific analysis can therefore be compared to mathematics, which also leads to pronouncements of absolute certainty, but we purchase this certainty at the price of the loss of all *concrete significance.* The object of the method of scientific analysis is the system of our science: its results are, to be sure, certain in the eyes of experimental science because they contain only the relations within the system of this science.[25]

It seems to me that this philosophical point of view is the only one that allows us to notice the true sense of the work of Einstein. The philosopher can offer a place of honor to the physicist in the history of philosophy because he has, through a reliable instinct, given to physics that freedom of which the philosopher is deprived. It should never be forgotten that this physicist was always fully aware of that which he created, that he himself was always cognizant of the philosophical meaning of his discoveries. This point can be seen in his eulogy of Ernst Mach where we find Einstein's remarks concerning the similar work of Mach.

25 Remark: this is an empirical claim, sociological or psychological, about the scientist, of our day. The assertions of the method of scientific analysis, as much as they are *certain*, are not only assertions about the science of *our day*, but about the logical structure of all science resting upon definitions. If the theory of relativity can attack and contradict philosophical propositions, it is because these propositions have the pretension to be enunciations about *reality.* All combinations of general principles of knowledge, as soon as they lay claim to a certain validity, constitute an enunciation that touches the real because it signifies that knowledge is possible in the framework of these principles.

"The ideas which have shown themselves to be useful in organizing things have been taken by us with such authority that we forget their worldly origins and regard them as given and immutable. They are then distinguished as 'logical necessities', as 'given *a priori*.' The way to scientific progress is rendered impractical for long amounts of time by such errors. It is therefore by no means an exercise for trivial amusement to analyze the ideas that have been accepted for so long, to show under what conditions they may be justified and used, how little by little they come from what is given in experience. Their authority is lost through their excessiveness, they are eliminated if they cannot be regularly justified, they are corrected if their correlation with the things given is too faulty, they are replaced by others if it is possible to establish a new system which seems for one reason or another preferable. Such analyses seem more often superfluous, diffuse, even themselves ridiculous to the specialists whose gaze is carried more willingly to the details. But the situation changes if the habitually used concepts are replaced by more precise concepts, under the pressure of the development of the relevant science. At that moment, that which has not proceeded correctly towards this concept gives rise to an energetic protest and causes a revolutionary overthrow of that which is most sacred. In such crises are heard the voices of those philosophers who believe that this idea cannot be abandoned because in it is the treasure of their 'absolute', of their '*a priori*', in short, because they built up so high that which they proclaim immutable in principle. The reader may again doubt that I am alluding to the doctrine of space and time as well as mechanics that has been modified by the theory of relativity. No one can deny that the inventors of the theory of knowledge have opened the path to these discoveries. As to my personal account the very least I know is that I have greatly profited, directly and indirectly, above all from Hume and Mach."

There is a small piece by Einstein on "Geometry and Experience" which formulates the relation between geometry and physics with a classical conclusion. To the question of the apodictic certainty of the geometric axioms, Einstein responds: "As far as the principles of mathematics refer to reality, they are not certain; as far as they are certain, they do not refer to reality." This formula can be extended to all propositions in the theory of knowledge. Philosophical considerations of this sort are not for Einstein the byproduct of his work in physics. Einstein says, "I attach special importance to the view of geometry which I have just set forth, because without it I should have been unable to formulate the theory of relativity. . . . The decisive step in the transition to general co-variant equations would certainly not have been taken if the above interpretation had not served as a stepping-stone." To understand the meaning of this remark, you must know Einstein's intuitive and extensive mode of thought through personal discussions. Einstein is not the formal mathematician for whom the mathematical symbols develop spontaneously into a theory. He thinks deeply, plunging himself into the essence of his ideas. Mathematics, for him, is only a means of expressing an intuitive process in which the laws come from unknown sources and in which the language of formulae serves only as a framework. It is a rare gift of destiny that this philosophical vision and mathematico-physical certainty have been united in the same brain. It required such creative genius to produce the theory of relativity. Here again let us remark that the course of things is not always determined by that which is believed to officially guide them. In our times, it has been the physicists and mathematicians who make the philosophy, not the official philosophers. It seems to be the nature of the human spirit that the handling of particular concrete problems brings closer together the enigma of knowledge with all the abstract reflection on the inconceivable phenomena of its very conception.

11

Planet Clocks and Einsteinian Simultaneity

An investigation of the extent to which astronomical measurements of the speed of light from the eclipsing of Jupiter's moons confirm the principle of the constancy of the speed of light. The result is a reduction of the question to the problem of absolute transport time.

Having placed the empirical foundation of the principles of the constancy of the speed of light in axiomatic form[1] and recognizing that several of these axioms have not yet received conclusive experimental support, it will now be of interest to consider an experiment to confirm the light principle which I had not mentioned, but to which Born has referred.[2]

The eclipses of the moons of Jupiter may be used for an astronomical measurement of the speed of light. It is well known that

Translated from "*Planetenuhr und Einsteinsche Gleichzeitkeit,*" *Zeitschrift für Physik,* vol. 33, no. 8 (1925), pp. 628–34.

[1] Reichenbach, *Axiomatik der relativischen Raum-Zeit-Lehre,* Braunschweig, Friedr. & Sohn Akt.-Ges., 1924.

[2] Born, *Die Relativitätstheorie Einsteins,* 3. Aufl. 1922, p. 100 and p. 191.

the delays in the eclipses of a moon will progressively increase over the course of a year; the resulting overall delay corresponds to the time that the light has taken to traverse the length of the axis of the Earth's orbit. Hence, we are measuring the speed of light in a *one* directional sense. Now Maxwell has already pointed out that the speed of light must be different depending upon whether it travels with or against the direction of the orbit of the Earth because the inertial system *S*, in which the sun and the elliptical orbits of the planets are at rest, will itself have a velocity *V* (which Born calls *v*) with respect to a preferred inertial system *J*, which according to the older theory is that in which the aether is at rest; the speed of light must therefore really be $c + V$ in one direction and $c - V$ in the other. These measurements are carried out individually; one is taken when Jupiter is at one end of its orbit and is repeated after half a revolution (six years), so that the light now traverses the Earth's orbit in the opposite direction. It has, however, yielded no difference in the value of the speed of light, thereby agreeing with the relativistic view in which the system *J* does not exist and in which the speed of light in *S* must equal *c* in every direction. Now Born correctly points out that these observations are not exact enough to serve as a confirmation of the relativistic view; in favor of the classical theory one may always make the assumption that *V* is too small to produce a noticeable effect. If *V* lies within the normal range of astronomical velocities, the effect will be unobservable. Therefore these observations will not be able to provide a real decision for the time being. It is worthwhile, however, to think through the question carefully and investigate what really would be demonstrated if the observations were exact. The value of such an investigation is in uncovering the relation between theoretical hypotheses and facts, thereby saving us from overemphasizing the experimental nature of the facts.

We therefore want to make the following assumption:

Axiom G. It is known from other observations that V is larger than some limit g and that the measurements of the speed of light from two diametrically opposed positions of Jupiter have shown that the difference between the resulting values is much smaller than g, hence we are not able to interpret them as $c + V$ and $c - V$.

What would this confirm?

It is first of all noticeable that velocity here is measured in a *one*-directional sense; this requires comparing time measurements at both ends of the Earth's orbit. But that is not possible if simultaneity is not already defined. Which definition of simultaneity is employed in these measurements? It must therefore be noted how time is measured. Here we measure it with clocks that are fixed on the Earth and thereby transported along with it; it is actually a question of the definition of simultaneity by the transport of clocks. And since the astronomical clocks are set according to the rotation of the Earth, the Earth itself is a transported clock carrying its synchronization with it.

In order to see this clearly, we can interpret the measurement operation in the following schematic way. Let t_0, the moment when Jupiter is closest to Earth, serve as a zero point for the Jupiter clock, and for the Earth clock the moment T_0 is the time when a signal sent from Jupiter at t_0 arrives at Earth. The moment t_1, at which the Earth is at the other end of its orbit, is now measured on the Earth clock; on the Jupiter clock let T_1 be the moment at which a signal is sent to Earth and t_1 the time when it arrives. The difference $\Delta = (t_1 - t_0) - (T_1 - T_0)$ is the time that the light took to traverse the Earth's orbit (the time to travel the path from

Jupiter to the beginning of the Earth's orbit is eliminated); this difference Δ is exactly equal to the sum of the delays of the actual *appearance* of the eclipse of the moon. If the Earth were to remain at its starting point, then it would be the case that $\Delta = 0$; that is to say the T-time of the Jupiter clock would be standardized by measuring the duration of the moon's period by the Earth clock where the Earth is moving perpendicular and the distance from Jupiter does not change. Let the time $t_2 - \Delta$, which a light ray from Jupiter would take to reach the starting point of the Earth's orbit, be displayed by a second Earth clock which would remain permanently at its starting point. The time that it takes for the light to traverse the Earth's orbit as measured by the difference of two clocks whose synchronization is established by transport, i.e., in the beginning the Earth clocks stood next to each other and were adjusted through a local comparison, then one would be taken to the other end of the Earth's orbit and would be considered to be synchronized on its path. This is therefore a question of transported synchronization, where the Earth is the transported clock.

Jupiter would also be used here as a clock, Born correctly points out; but the more important clock in the problem is the Earth because the transport of the simultaneity of the Earth clock is a crucially important factor. Why should the speeds of light be different in the two directions? That is simply a question of simultaneity. Simultaneity for system S *can* be defined in such a way that the speed of light will be equal in both directions. That is no problem in this case. In order for such a definition to be univocal and free from contradiction we need only assume certain undisputed facts about the movement of light given by classical optics (axioms I–IV);[3] the Einsteinian definition of simultaneity,

[3] See Reichenbach, loc. cit., §21.

which is given by this definition, is certainly realized in S. The only question therefore is *whether the simultaneity transported by the Earth clock coincides with Einstein's simultaneity in the system S, or whether it will be identical with the simultaneity of the hypothetical system J.* If the difference in the speed of light $= 2V$, i.e., the individual speeds can be interpreted as $c + V$ and $c - V$ respectively, then the latter conjecture would be confirmed; equality of the speeds would confirm the first conjecture.

This question can only be settled through observation. It is not the question of the speed of light, but rather of the *absolute time of transport* that is the point at issue. If axiom G is confirmed, then the absolute time of transport (formulated as axiom B, op. cit.) is refuted, and we would gain an important *negative* result in support of the theory of relativity. We will investigate the *positive* results that derive from it.

There is regrettably not much to be gained here. This is because we must keep in mind that in a strict observation, agreement between the transported simultaneity and the Einsteinian is not possible, particularly if the theory of relativity is assumed. The Earth clock moves with a speed of v relative to S and therefore endures a deceleration relative to the Einsteinian time of S. If the astronomical measurements of the speed of light were absolutely exact, then they would confirm the time dilation $\sqrt{1 - (v^2/c^2)}$ in deviation of the value of c compared to the terrestrial measurements; but this deviation is of course within the experimental error at present. That is why the astronomical measurements cannot say anything quantitative about the time dilation factor at present, which now is understood only qualitatively. In order to understand this we must yet come closer to the significance that synchronization by transport has in the theory of relativity.

Indeed, this theory teaches us that synchronization by transport deviates from Einsteinian synchronization. But this deviation depends upon the velocity of transport v, and it has been calculated that the transport synchronization by small v increasingly approximates that of Einstein. It is a positive statement; it says, for the boundary case of infinitely slow transport velocity we get the Einsteinian synchronization for the system in question, and not at all an absolute synchronization. That is completely consistent with the theory of relativity because it confirms the *relativity of transport synchronization* if it is at all possible to give it a univocal definition.[4]

We can formulate this statement in a stronger fashion. Let $\varphi(v)$ be the time dilation factor of the clocks, i.e., $\Delta t' = \Delta t \cdot \varphi(v)$; $\Delta t'$ here is the specified time of the system in which the clock is at rest, and therefore identical with the time on the clock. Consider two distinct points, P_1 and P_2 in S, at a distance x from each other and a clock transported from P_1 to P_2; if it was synchronized with a stationary clock at P_1, it will display a difference with a stationary clock at P_2 of

$$\delta = \Delta t' - \Delta t = \frac{x}{v}\left[\varphi(v) - 1\right],$$

since $\Delta t = \frac{x}{v}$. For the Lorentz transformation is $\varphi(v) = \sqrt{1 - (v^2/c^2)}$ which is approximately $= 1 - \frac{1}{2}\frac{v^2}{c^2}$, therefore for small v,

$$\delta = -\frac{1}{2}\frac{xv}{c^2}.$$

The negative sign means that the moved clock runs slow.

[4] This has been pointed out by J. Winternitz in *Relativitätstheorie und Erkenntnislehre* (Leipzig: B. G. Teubner, 1923), p. 83.

We now compute this amount assuming absolute transport time. We therefore assume that a transported clock always runs on the time t^* of the privileged system J; despite that, we use Einstein's definition of simultaneity in S and ask how big the deviation δ will be relative to S-time under these assumptions for a transported clock with a speed v. For the calculation we use the transformation formulas I have derived elsewhere;[5] they are for the time coordinate:

$$t^* = t + \frac{Vx}{cc'}$$

or if we set $c = c'$, i.e., we posit an additional axiom (axiom X, op. cit., p. 81),

$$t^* = t + \frac{Vx}{c^2}.$$

By assumption, the time t^* is displayed on the transported clock; it therefore does not depend at all on v, and the time dilation factor $\varphi(v)$ amounts to

$$\varphi(v) = \frac{\Delta t^*}{\Delta t} = 1 + \frac{1}{\Delta t} \cdot \frac{Vx}{c^2} = 1 + \frac{Vv}{c^2},$$

where again $\Delta t = \frac{x}{v}$. Thereby, we will get

$$\delta = \frac{x}{v}\left[\varphi(v) - 1\right] = +\frac{xV}{c^2}.$$

The positive sign indicates that the clock is fast; but this is irrelevant. It is much more important that δ does not depend upon v; the synchronization of the transported clock in S will therefore not be influenced at all by speed of transport relative to S, but is merely a function of the speed V of the system S relative to the privileged system J. Consequently, δ does not go to zero along with v,

[5] Op. cit., p. 79 (64).

contrary to the previously computed case; hence, also in the case of infinitely slow transport speed there is a deviation of transport synchronization from Einsteinian synchronization for *S*.

If the axiom G is confirmed, then this is evidence that the formula $\delta = xV/c^2$ is false; because to be able to distinguish between the values $c + V$ and $c - V$, this amount must lie outside of the limit of experimental error. But this then also demonstrates that the clock that moves relative to *J* with the resultant speed of speed *V* combined with *v* does not take on the time of *J*; viz., it proves the existence of a time dilation effect. If we further assume that *v* is small relative to *V*, and that the amount $\frac{1}{2}\frac{xv}{c^2}$ is within the limit of experimental error in the measurement of *c*, then by means of axiom G the following theorem is confirmed:

Theorem G′: There exists a time dilation with the measurement factor $\varphi(v)$, such that for points P_1 and P_2 with a fixed distance *x*, the difference δ between a clock transported from P_1 to P_2 with a velocity *v* and a clock at rest at P_2, i.e., the function $\delta(x, v) = (x/v)[\varphi(v) - 1]$, goes to zero as *v* decreases if *v* is measured in an inertial system according to the Einsteinian synchronization.

It is the last part of this theorem that is important because the special role of Einsteinian synchronization rests upon it. It must further be noted that the difference with the absolute transport time rests on the fact that the function $(x/v)[\varphi(v) - 1]$ goes to zero with *v*. Contrarily, $[\varphi(v) - 1]$ goes to zero with *v* in both cases. This clearly shows that by keeping the amount of time constant, the deviation between a moving and a stationary clock goes to zero with *v* for both Einsteinian and absolute time; by keeping the *spatial* distance constant, however, this deviation only goes to zero for the Einsteinian time (except in the privileged system *J*).

Theorem G′ would therefore be the entire profit we receive from greater exactness in the determination of the velocity of light by Olaf Römer's method, i.e., from a confirmation of axiom G. Certainly this would be a significant result in that it refutes absolute transport time and proves Einsteinian simultaneity to be the privileged simultaneity for every inertial system. But to tell the truth, the form of the factor $\varphi(v)$ that is important for the problem of clocks would not be clarified.

A confirmation of the theorem of the constancy of the speed of light cannot be asserted. The content of this theorem about the relation between the speed of light, time units, and measuring rods will not be affected by the aforementioned measurements; they will not be confirmed until the axioms IX, X, and D receive experimental confirmation, as I have mentioned elsewhere.[6] The astronomical measurement of the speed of light can only prove the privileged place of the Einsteinian definition of simultaneity as confirmed by the use of planet clocks. With it, however, the theorem of the constancy of the speed of light is not more true; because Einsteinian simultaneity is a valid definition, it is beyond question and does not need confirmation. In the above-mentioned distinction, an indication can be seen of the actual existence of the relation between clocks and light conjectured by Einstein.

[6] Op. cit., pp. 73 and 88. Two of these axioms would be sufficient.

12

On the Physical Consequences of the Axiomatization of Relativity

I. What are the actual physical statements of the theory of relativity, aside from the general theoretical view? II. Response to some objections. III. The experimentally unconfirmed assumptions of Einstein's space-time theory. IV. Lorentz contraction, Einstein contraction, and their confusion.

I.

Having already given a comprehensive account of an axiomatization of the theory of relativity,[1] I would now like to expand upon those consequences which are especially important for physics. While this was originally intended to be a work of epistemology, and its real significance is primarily in that field, it does, however, provide some important results for the experimental foundation of the theory of relativity.

Translated from "*Über die physikalische Konsequenzen der relativischen Axiomatik,*" *Physikalische Zeitschrift,* vol. 34, no. 1 (1925), pp. 32–48.

[1] *Axiomatik der relativistichen Raum-Zeit-Lehre* (Braunschweig: F. Vieweg, 1924). Hereafter to be referred to as A.

There are two approaches to physical axiomatization. In formulating a *deductive axiomatization*, the most general principle possible, perhaps a variational principle, is placed at the top and all other details are derived from it; only these derived details are testable, and if they are verified we may consider the abstract axiom to be more or less probable. *Constructive axiomatization*, on the other hand, proceeds differently. Here we take as axioms only those statements that are themselves direct experimental results; from them we derive the entire theory by integrating some additional conceptual elements, viz., definitions. The definitions are arbitrary and therefore can never introduce error into the theory. This form of axiomatization has the great advantage for physics that the *implications of each experimental result* can be immediately recognized. Not every principle of the theory pre-determines *every* axiom, and it can therefore be immediately known, as soon as such a structure is put in place, whether a principle is based upon well-grounded axioms or if it presupposes uncertain axioms.

Some have objected to my use of the term "axiom" for statements confirmed by experience. But it is standard practice in physics to refer to such propositions as axioms. A proposition is called an axiom of a physical theory if it serves as a logical meta-principle of that theory; but it is thereby liable to the same empirical objections as the theory itself. *The "most foundational observation statements" we call physical axioms.* On the logical status of definitions, please see §2 of A. where the peculiar character of physical definitions is described in terms of *coordinative definitions.*

Since my axiomatization is constructive in that sense, it offers physicists the great advantage of being able to immediately recognize the experimental guarantee of the individual propositions of the theory of relativity. The need for such a strict construction is unfortunately not widely acknowledged. One usually finds a peculiar belief on the part of physicists: when the author of a theory employs theoretical considerations as a starting point in explaining some given observable effects, it is then held that these very effects in turn confirm the original theoretical considerations. But the unbiased critic will not rest content with such a historical demonstration, he will demand the logical connections between the observational data and the theoretical foundation. For this reason, such a representation is relevant not only to philosophy, but to physics as well because in introducing so foundational a change into the physical worldview, the physicist cannot be careful enough with the conclusions he draws. He will at the same time succeed in finding a representation of the theory that is strictly correct, and thereby conceptually coherent, and which will free his conscience from the several leaps which would otherwise defy the logical shoals laying in the foundation of the theory.

The axiomatization is constructed in the following manner. Consider space to be filled with point masses similar to the chaotic distribution of molecules in a gas with an observer sitting on each point mass. The observers can communicate through light signals. The task is then to select one particular system of such point masses that are, on the basis of certain properties, called "at relative rest," and in this "rigid" system we define a space-time metric. Those facts concerning the movement of light that are thereby required are termed *light axioms*; these axioms, which can be formulated without the concept of simultaneity, are entirely free of statements concerning material objects such as measuring rods and clocks. The completion of this task generates a completely feasible *light geometry* when we add to the light axioms the definitions of simultaneity, the equivalence of time intervals, the mutual rest of points, equivalence of length intervals, etc. *Within* any point system, the light geometry is *univocal*, but the light geometry does not univocally determine the collection J of inertial systems – it only determines the more general collection S of systems related by four-dimensional similarity transformations. S contains J as a subgroup. This subgroup can be defined by considering rigid bodies or natural clocks; only the systems J have the property that their points can be connected by rigid rods or that the light clock defined by two points will measure the same duration as a natural clock.

There is a mistake in this respect in A. that is not essential for the completed construction, but which I would like to correct on this occasion. The distinction between the classes S and J is correctly drawn; but the conclusions based upon the light geometry are the result of a mathematical mistake. Let the class T be the part of S that does not belong to J, symbolically, $J + T = S$. In A. the point is presented: a system J and a system T cannot *permanently* possess a common space point. Unfortunately, this proposition is false.

It leads immediately to the following thesis: Select any point A as a starting point, the system S constructed around it according to the rules of the light geometry is *univocally* determined; it is possible to obtain only one of the systems J or T depending upon the state of motion of A. This is the form in which I have proved the proposition, but I am greatly obligated to thank Hans v. Neumann, Zurich, for discovering a mistake contained in the proof. Equation (21) on p. 45 in A. is incorrect because the coefficients of the terms of the second order are evaluated, while (19) is only valid to the first order. In this way, the proof is unsound. In the same way, the second proof of the theorem (A., p. 47) given by K. Fladt contains a mistake: the formula following equation (31) is wrong, here the numerator and denominator of the fraction on the right-hand side do not have to be individually independent of η_1. Because of this error, the general solution for the theorem of fig. 7 of A. is not the linear function Aτ + B, but the rational function $(A\tau + B)/(C\tau + D)$; a corrected proof of this was developed by v. Neumann as an extension of Fladt's proof. Consequently, my presentation in §16, in which the systems *T* are considered, must be inserted at this point. It is therefore the case, contrary to the assertion on p. 62 of A., that by maintaining the starting point, more than one system can be constructed around it such that the light axioms are valid.

The resulting light geometry is *arbitrary* because of the additional definitions. Depending on the definitions chosen, the resulting light geometry may be relativistic or classical. The relativistic light geometry therefore only differs from the classical in these definitions, since either the Lorentz or the Galilean transformations can be defined. The only new aspect of the light axioms in the theory of relativity is the consideration of the speed of light as a limiting velocity; but even if this axiom holds, the Galilean transformations can still be defined. (For velocities $> c$ "real systems" are certainly no longer constructible.)

A second class of axioms, the *matter axioms*, comes after the light axioms covering the behavior of clocks and measuring rods. First, the concepts of "rigid rod" and "natural clock" must be defined independently of the geometry that results from them. It has been occasionally objected that rigid bodies can only be defined through an appeal to Euclidean geometry, but this is incorrect. They can also be defined without an appeal to the geometry if one makes use of the distinction between metrical and physical forces; see §18 in A.

The content of the matter axioms may now be summarized in the following manner: *material objects obey the relativistic light geometry.* Distances that are light-geometrically equal are of equal length when measured with a rigid rod. Light-geometrically equal durations will also be equal when measured with a natural clock. We must also consider statements concerning the transport of rods and clocks into moving systems; agreement here arises as well when they are compared to units that have been transferred light-geometrically onto a system of motion. Of course, we must use the definitions of the relativistic light geometry, not the classical ones. Einstein's assertion can thereby be plainly expressed as "material things obey not the classical, but the relativistic light geometry."

We are now at the point where we are facing that which is physically novel about the theory of relativity. While the light axioms are already valid in classical optics, the theory of relativity only adds the proposition that the speed of light is an upper bound for signal velocities; in this way the matter axioms embody a deviation from the classical theory. They contain the assertion that the Lorentz transformation, which differs from the Galilean transformation only *in terms of definitions*, is in fact the transformation for rods and clocks. This assertion contains the portion of the relativistic theory of space-time that is experimentally verifiable, so it is only this assertion that is open to physical discussion. It is senseless that again and again the philosophical principles of the theory of relativity are disputed, for these are now secured beyond all doubt. It is also pointless to seek contradictions in the relativistic system, since it clearly follows from axiomatic construction that the axioms and definitions of the system are *logically possible.* Each individual axiom signifies a conceivable matter of fact in which there is absolutely nothing mysterious or unimaginable. But whether the axioms are *true*, precisely in the sense of true that we mean for every other physical proposition, and therefore whether they are in agreement with experience, is what is under investigation. While a number of the axioms have been confirmed, others are yet to be. We will go into more detail on the unconfirmed axioms in III.

With the aid of the axiomatization, the claims of the theory of relativity can be strictly formulated; the claims we are making here are simply laid out without cloaking them in the compound "relativity principle" and "principle of the constancy of the speed of light." While it was extremely fruitful for Einstein to lean on these two principles while he was concerned with creating his theory – at that point the concern was to formulate the basic idea in such a general way that one could almost guess the concealed intentions of nature – now, for the logical investigation of the theory, we must place the narrowest and most basic individual assertions at the top. At the same time, it succeeds in separating out the place of experience in the theory from the concepts that needed to be added to it. By separating axioms from definitions, we are able to distinguish between those propositions concerning the motion of light which speak of its physical characteristics and those arbitrary additions, like the concept of simultaneity; and we are in a position to be able to univocally describe the behavior of material objects without needing any ambiguous auxiliary concepts like "shortening" or "stretching." On the basis of our axiomatic presentation, we can therefore finally pick out *that which the relativistic space-time theory maintains about reality.* The discovery of these assertions from the mathematically splendid, but epistemologically opaque, construction of a physically geometric formulation of the theory of relativity is the true achievement of this axiomatization.

II.

Unfortunately, the described intention of my investigation was misinterpreted by some. Weyl, for example, has taken the view that the goal of my investigation is essentially mathematical. He writes,

> The central focus of the book is not philosophical, but rather a purely mathematical investigation and thereby it must be judged from a mathematical point of view. However, in this regard it is quite unsatisfactory being too cumbersome and overly obscure. The main point, the axioms a, b, and c (the light axioms) and transition to space-time measurement, to coordinate space, and to Möbius geometry could have been conveniently accomplished in a couple of pages thereby gaining clarity and intelligibility.[2]

I must strenuously defend my axiomatization against this objection. From the mathematical angle, the only options are for my work to be judged as "true" or "false." Mathematical elegance was not my aim – relativity theory itself provides sufficient opportunity for that, and in any case it has limited value in epistemological discussions. In the opening pages of his review, Weyl gives a presentation of my axiomatization that is supposed to serve as an illustration of how I could have done it "conveniently in a couple of pages"; I gladly leave it to the judgment of the reader to determine which presentation is less "cumbersome and obscure." For my part, I have always preferred the clarity that comes from a gradual construction from the simplest possible logical operations to the way in which some obfuscate their thoughts with a scintillating mathematical fog. The plan of my investigation is

[2] *Deutsche Literaturzeitung*, Nov. 1924, p. 2122.

guided by the intention to bring out the results of physical experience as clearly as possible and to uncover as many conclusions as can be derived from each new observation statement. When working with a minimum of concepts, some steps do seem more complicated when compared to treatments that begin with the entire set of available resources laid out before you. But I find it extremely regrettable for a mathematician of Weyl's standing to so misunderstand the aim of such an epistemologico-logical investigation and to use his authority in this attempt to suppress the establishment, at long last, of a logical foundation which, in the end, can alone guarantee the validity of the theory of relativity in all its mathematical and physical fertility.

What needed to be done from the mathematical direction was completely carried out by Carathéodory,[3] whose representation was written without knowledge of my investigation (in a survey already published in 1921). The fact that he saw fit to follow essentially the same path for purely mathematical reasons provides strong support for the logical justification that stands behind the epistemological grounds of my construction.

In this context, I must briefly mention the criticism of my book proposed by A. Müller.[4] He calls my theory of definitions "not very well thought out" and wants to say that if one reads between the lines in my book that you will find the view that I deny the factual nature of the theory of relativity. This is such a radical misunderstanding that I must simply reject any fault on the part of my account. Although each individual coordinative definition is neither true nor false, the theory of relativity, when interpreted through the doctrine of the arbitrariness of these definitions, can

3 *Zur Axiomatik der speziellen Relativitätstheorie.* Sitzungsber. d. Berl. Akad., 1924.
4 *Phys. ZS.* 25, p. 463, 1924.

make a decisive claim to truth. This basic logical distinction is obviously contained in my presentation. I address Müller's other objections elsewhere.*

III.

I now turn to a more detailed discussion of those axioms in need of greater experimental support.

The light axioms (I–V in A.) can be considered to be well confirmed. Among them, the only new fact is the limiting nature of the speed of light about which surely there is no longer any serious doubt.

Amongst the matter axioms, VI and VII only contain the long-held assertions about clocks and rigid rods that also follow from the classical theory. Since axiom VIII is a formulation of the Michelson experiment, it is thereby considered well confirmed. Further, the result has received long-awaited confirmation by the experiments of Tomaschek using the light emitted by stars, which were inspired by the work of Lenard.[5] Recently, doubts have been raised by Dayton C. Miller,[6] who obtained a positive result on Mount Wilson; yet to be determined are the cause of this result and what should be inferred from it. In this context, the axiomatization is proved to be extremely useful because it shows what particular role the Michelson experiment plays in the theory, what follows from it, and what is independent of it. Because of the importance of these experiments we will consider them in detail in IV.

* See Chapter 9 in this collection.

5 *Ann. d. Phys.* 73, p. 105, 1924.

6 *Proc. Nat. Acad.* 11, p. 306, 1925, no. 6. This formulation of axiom X differs slightly from that in A. in which the constraint to inertial systems is not posited.

We now turn to the matter axioms that have never been experimentally proven. They are as follows:

Axiom X. *The time unit of a natural clock is always of such a kind that the time ABA of a light signal measured by these clocks at A is the same in all inertial systems if AB is the same when measured by rigid rods.*

Axiom IX. *The unit transported by a rigid rod transported from one inertial system K to another K′ has the following property: the length of the unit at rest in K as measured from K′ is equal to length of a unit at rest in K′ when measured from K. And this holds when Einstein's definition of simultaneity is applied to both K and K′.*

Axiom D. *A clock moving with a uniform velocity v relative to a certain system K, to which the Einstein's definition of simultaneity is applied, experiences a retardation to the degree* $\beta = \sqrt{1 - (v^2/c^2)}$.

These three axioms are not mutually independent. When two of them are confirmed, the third will follow from them given the confirmation of the rest of the system. We will look at each of these three axioms individually.

Axiom X is an important part of the Principle of the Constancy of the Speed of Light. This principle is based upon several axioms and one definition, the definition of simultaneity. It can be understood in three senses. Assume only the light axioms, the light principle says that the light geometry can be set up so that the speed of light is constant. If we add the Michelson experiment (and axioms VI and VII), then we can obtain the broader result that the speed of light is also constant when spatial distances are measured with rigid rods. The third and broadest meaning, however, follows from axiom X, which says that this constant has the *same numerical value* in all inertial systems if measured with natural clocks and rigid rods that possess equal rest units.

However, there have been no attempts to confirm this axiom. The Michelson experiment proves nothing in this respect. This results from the fact that the system of axioms I–VIII, which contains the Michelson experiment and the negation of axiom X, can nevertheless be combined to form a consistent system (§23 in A.).

Axiom X has an intuitive meaning if the light-emitting atom is considered to be a clock. It then asserts that the connection between the frequency of light, the speed of light, and measuring rods is the same in all inertial systems. Now the Earth at different points in its elliptical orbit occupies different inertial systems; this axiom thereby asserts that the wavelength of a given spectral line will be the same in all seasons.

To fully understand this fact, we must make an additional comment. Wavelength is a concept that is dependent upon the definition of simultaneity, since what we call wavelength is the distance between the equivalent phases of a wave at the same time; it thereby also depends upon the speed of the wave in *one* direction. One additional proposition is therefore needed here, that in the various inertial systems Einstein's definition of simultaneity will always be used in measuring wavelengths. But it is not necessary to observationally confirm this additional proposition in practice, since it is already satisfied when one measures *standing* waves. The length of a standing wave is independent of the definition of simultaneity. Consider a light ray sent from A to B and then reflected back to A. If in accordance with Einstein, I place the time of arrival at B at the midpoint of the time of the round trip

ABA as measured from *A*, then the wavelength is the same in both directions and equal to that of a standing wave. If I place the time of arrival at *B* differently, by attributing to it a different value in the interval *ABA*, then the wavelengths will be different for different directions but so will the velocity; since standing waves that arise through interference, again, as one can quite easily see, have the same wavelength as obtained from the first definition. The length of a standing wave therefore does not depend upon the definition of simultaneity; the node points represent something truly objectively observable, and their distance can be measured without a definition of simultaneity.

Consequently axiom X can be confirmed if one very precisely measures the wavelengths of a specific spectral line in different seasons. Unfortunately, the current degree of accuracy is not sufficient. A precision of at least 10^{-8} in measuring the wavelength is necessary to check it, while currently we can only achieve an accuracy 2×10^{-7} (Michelson).

We now come to axiom IX. This axiom implies a definite distinction for Einstein's simultaneity; it says that there is symmetry between two systems using Einstein's simultaneity with respect to measurement of length and this can therefore be called the requirement of the "relativity of measured length." It is equivalent to the assumption that in the Lorentz transformation, orthogonal to the direction of motion, the coordinates transform identically; this axiom can be derived from this assumption if the confirmed axioms I–VIII are added. (Conversely, this assumption can be derived from axiom IX.) Because of this connection, the axiom is very plausible. But it seems at present that there is no way to directly prove the axiom.

The situation is much better with axiom D. This axiom also distinguishes Einstein's simultaneity because the velocity v of the time dilation factors is only defined if simultaneity is defined in *K*. It furthermore seems to distinguish a system *K*; this is however only an appearance because it must be pointed out that the time dilation factor is the same in *every* inertial system if it is given for *one*. (This follows from axioms I–VIII and D; see A., p. 87.) Consequently, axiom D is equivalent to an axiom E which is a formulation of the "relativity of time measurement" as an analogue to axiom IX. Nevertheless, we choose the less elegant form of axiom D because this axiom corresponds to an immediately verifiable fact: it is a formulation of the *transverse Doppler effect*.

The calculation of the amount of the Doppler shift by Einstein is different from that of classical kinematics because, contrary to the classical Doppler effect, Einstein's would be affected by the time dilation of the Lorentz transformation. But this difference is not measurable because it can always be attributed to the velocity of the source, which cannot be directly measured with exactness. On the other hand, the difference becomes barely measurable if the source moves *transversally*, i.e., vertically with respect to the line that connects the source and the observer. According to Einstein, however, a (very small) Doppler effect in the sense of a decrease in frequency arises, whereas in classical kinematics there would be none in this case. This *transverse Doppler effect* is identical with the time dilation effect of the Lorentz transformation as we have formulated it in axiom D.

The effect becomes barely measurable if canal ray particles are used as light-emitting clocks. We need not measure their velocity, but rather only their *direction*. The zero point *N* of the Doppler effect (i.e., equal frequency for gas molecules at rest and in motion) then does not lie at an angle of 90° to the direction of motion (see Fig. 12.1, where *O* designates the point of emission of light), but in the sense of the direction of motion having been shifted by an angle δ at the particular place where Einstein's time dilation is compensated by the Doppler effect that causes a small increase in frequency. This difference is calculated to be $\delta = (v/c)/2$ where v represents the speed of the canal rays and c is the speed of light.[7] Thereby, $\delta = 11\frac{1}{2}$ arc minutes if we assume the speed of the canal rays to be $v = 2 \times 10^8$ cm/s. This displacement can be measured most accurately if one considers the zero points on both sides of the direction of motion and measures the displacement 2δ of the angles 2θ from 180°. Even a solely *qualitative* confirmation of the deviation from 180° would be a success because even though axiom D would be left unconfirmed, *absolute transport time*, which stands opposed to time dilation, would thereby be refuted.

At this point, however, no experiments of this sort are currently under way; they would be greatly in the interest of the theory of relativity.

In summary, we can say: of the three unconfirmed axioms X, IX, and D, D has the greatest prospects for quick experimental verification and IX, the least. With the confirmation of D we do not confirm the Lorentz transformation, but only guarantee a slightly more general transformation (the transformation 75 in A., p. 87). After D is confirmed, a certain more general viewpoint would support IX, since, as demonstrated above, we can show that the

7 Elementary derivation: The time dilation factor is $= \sqrt{1 - (v^2/c^2)}$, approximately $= 1 - (1/2)(v^2/c^2)$; the difference in units between moving and resting clocks is therefore $= (1/2)(v^2/c^2)$. This is compensated by a Doppler effect that is caused by the component $v \sin \delta$ of the velocity, and therefore amounts to $(v \sin \delta/c)$. The above-mentioned value follows as a result of equating both expressions and substituting δ for $\sin \delta$. – The strict theory gives the same for small v/c.

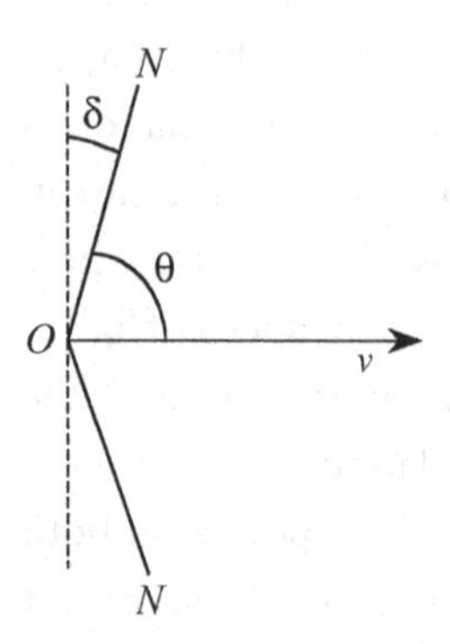

Figure 12.1.

relativity of time intervals is already satisfied by this more general transformation; and it could be supposed that it would then also hold for length contraction. The complete Lorentz transformation would be established up to the certainty of this supposition, since from D and IX, X and thereby the complete Lorentz transformation follows when one adds the confirmed axioms I–VIII.

The point of this discussion should in no way be taken to be that I regard the confirmation of these axioms as unlikely. To the contrary, they seem to me, given the current situation, to be more likely than the axiom of absolute transport time;[8] these axioms fit better in the system of the already-confirmed axioms than those of absolute time, especially with respect to axiom VIII, the result of the Michelson experiment. But it nevertheless seems important that the experimental physicist take note of these gaps in the theory of relativity in the hope that they can be filled in by experiment before too long.

[8] On the possibility that this time is refuted by astronomical measurements of the speed of light, see Reichenbach, "Planetenuhr und Einsteinsche Glecihzeitigkeit," *ZS. f. Phys.*, vol. 33, p. 628, 1925. [See Chapter 11 in this collection. Portions of this section were included by Reichenbach verbatim in *Philosophie Der Raum-Zeit Lehre.*]

IV.

In conclusion, we should examine a particular error that has crept into the understanding of the theory of relativity. It concerns the problem of Lorentz contraction and thereby leads us to the Michelson experiment.

One frequently hears the opinion expressed that in the Lorentzian explanation of the Michelson experiment the contraction of the arms of the apparatus is an "ad hoc hypothesis," whereas Einstein explains it in a most natural way, namely, as a result of the relativization of the concept of simultaneity. But this is false. The relativity of simultaneity has nothing to do with length contraction in the Michelson experiment, and Einstein's theory gives just as poor an explanation of it as that of Lorentz.

That this opinion is false already follows from the fact that the contraction of one of the arms of the apparatus occurs precisely in the system in which the apparatus is at rest. The "Einstein contraction" only explains that the arm is shortened if it is measured from a different system. But that does not explain the Michelson experiment, since it proves that the rod lying in the direction of motion is shorter *when measured in the rest system* than it should be according to the classical theory. If there were a privileged inertial system J with two rigid rods of equal length, one measured according to the classical theory and the other according to Einstein's theory, they would no longer be of equal length in an inertial system S if they would lie in the direction of motion; the Einsteinian rod would be shorter. In particular, this difference can be measured as a difference of "rest length" in S as well as a difference in the "length of a moving rod" as measured from a different inertial system. The Einsteinian theory, as well as Lorentz's, differs from the classical theory in asserting a measurably different effect on rigid rods that has nothing to do with the definition of simultaneity.

The concepts *real* and *apparent* are often brought into this discussion such that the Einsteinian contraction is considered to be apparent in contrast to the Lorentz contraction which is genuinely real. While these concepts are indeed confusing, it must be acknowledged that there is a distinction here that must be correctly formulated. It works in the following way. The Einsteinian length contraction that results from simultaneity is related to the comparison of *different magnitudes* within the *same theory*. The "length of moving rods" is entirely different from the "length of resting rods" just as the "perceived angles of objects at 10 meters" is different from the "perceived angle of objects at 100 meters"; both notions of length concern different magnitudes, and it is no wonder that they are assigned different values. This difference can be called a *disjunctive distinction*; it concerns different objects placed side by side. This state of affairs is obscured by the fact that these objects can be understood as distinct logical functions of the same argument (i.e., the rod); but they are *different* functions possessing the same status. This will be very clear if we interpret both notions of length as different particular cuts of the world-lines of the rods in Minkowski space-time; this is just like the above example of perceived angle.

But this is different from the true contraction of Lorentz which compares the behavior of the *same magnitudes* according to *different theories.* We therefore have distinct *claims of truth*; the two assertions are mutually exclusive, the *same* rigid rod behaves differently under the *same* type of length measurement depending upon whether Lorentz's or Einstein's theory is right or if the classical theory is right. We speak therefore of an *alternative distinction*; in this we compare the *real* behavior of things to a *possible* behavior. In other words: it is not the concepts of real and apparent that are brought into the discussion here, but the concepts alternative and disjunctive. And the alternative distinction is drawn between the explanations of the Michelson experiment given by Einstein's theory and the classical theory, and by Lorentz's theory and the classical theory, whereas there is no distinction of this sort to be drawn between Lorentz's and Einstein's theories; in particular, both assert the facts of the case set out in axiom VIII of A. (p. 69), where the classical theory claims the facts to be otherwise. The notion of simultaneity does not seem to be present at all in this situation.

One should, therefore, not call these "contractions" by the same name. On the one hand, there is Einstein's contraction that results from the relativity of simultaneity and compares the lengths of moving and resting rods. On the other hand is the Lorentz contraction that compares the length of a rigid rod according to the empirical facts of the Michelson experiment with the length of the rods determined by the classical theory. It just happens that both contractions depend upon the same measurement factor $\sqrt{1 - (v^2/c^2)}$, and this is probably the reason why they are always confused with one another. The two have completely different meanings. Both the Einstein contraction and the Lorentz contraction arise in Einstein's theory, and with respect to the Lorentz contract, neither Einstein nor Lorentz offer an explanation, but simply assume it axiomatically.

If we say that the two measurement factors "just happen" to be equal, then we are saying that their equality depends upon certain requirements; but of course there exists a theoretical relation between the two factors. It can be proven that they must be equal under the requirement of *the linearity of transformation*, but *only* under this requirement. Proof: Let l be a rod that behaves according to Lorentz's, i.e., Einstein's theory, and compare it to L which is a rod that behaves according to the classical theory; their rest lengths in K are equal, i.e., $l^K_K = L^K_K$ (the upper index refers to the system in which it is measured and the lower refers to the system in which the rod is at rest). The Lorentz contraction is related to the ratio:

$$l^{K'}_{K'} : L^{K'}_{K'}. \tag{1}$$

Einstein's contraction, on the other hand, is concerned with the ratio:

$$l^K_{K'} : l^K_K. \tag{2}$$

Now, in the classical theory $L^K_{K'} = L^K_K$ (in this comparison only the simultaneity of K is employed, that of K' is not used at all), therefore because of the first mentioned equality it is also true that $L^K_{K'} = l^K_K$. Hence, (2) is equivalent to the ratio:

$$l^K_{K'} : L^K_{K'}. \tag{3}$$

Because of the linearity of transformation (and only because of this) (3) is the same ratio as (1), thereby giving us (2) as the same ratio as (1).

If we now want to look at the question of the required *explanation*, we first have to point out that the problem is being extraordinarily obscured by the use of the word "contraction." This word seduces us into misusing the concept of causality. In seeking a *cause of the contraction*, we therefore believe that we must find a cause of the *difference* in the compared quantities. This idea has caused great confusion. It brings about the preference for one theory, the classical; *no* cause is required to explain why objects follow its laws, and causes are only needed when there is a *deviation* from the expected behavior. But it is obvious that the same causal problem arises if the measuring rods and clocks are adjusted according to either the classical or relativistic transformation. The word *adjustment*, first used in this way by Weyl, is a very good characterization of the problem. It cannot be a coincidence if two measuring rods placed next to each other are of the same length regardless of their location; it must be explained as an *adjustment to the field* in which the measuring rods are embedded as test bodies. Just as a compass needle adjusts to its immediately surrounding magnetic field by changing its *direction*, measuring rods and clocks adjust their *units of measure* to the *metric field*. All metrical relations between material objects, including the observed facts of the Michelson experiment, must therefore be explained in terms of the particular way in which rigid rods adjust to the movement of light. Of course, the answer can only arise from a detailed theory of matter about which we have not the least idea; it must explain why the accumulation of certain field loci of particular density, i.e., the electrons, express the metric of the surrounding field in a simple manner. The word "adjustment" here thus only means a *problem* without providing an *answer*; the relevant *fact* is strictly formulated in the matter axioms without using the word "adjustment." Once we have this theory of matter, we can explain the metrical behavior of material objects; but at present the explanation from Einstein's theory is as poor as Lorentz's or the classical terminology.

What then is the advantage of Einstein's theory over Lorentz's? It cannot be that Einstein gives an explanation of the Michelson experiment; it says nothing about the Michelson experiment except to include it as a basic axiom of the theory. However, it lies in the fact that Einstein's theory *foregoes* an explanation of the experiment in terms of a "contraction." This "explanation" from Lorentz' theory is its mistake; it assumes the classical relations to be "self-justifying" and posits the faulty causal claim that the deviation must be understood as being brought about by some cause.

Now we can also address the question what would change in the theory of relativity if Miller's experiment were held to prove that the hitherto negative result of the Michelson experiment is in principle wrong. *Nothing* would change in Einstein's *theory of time* as it has nothing to do with the Michelson experiment. Also nothing would change with the *light geometry*; it remains in any case a possible definition for the space-time metric and probably a much better and more accurate one than the geometry of rigid rods and natural clocks. But what would change is our knowledge about the adjustments of material things to the light geometry. With respect to the matter axioms, as far as they differ from the classical theory, the Michelson experiment is the only one that has been confirmed. If this should be refuted, one has to develop a more complex view of the relationship between material objects and the light geometry.

But it would not completely shake the foundation of relativistic physics. It is actually very unlikely at this point that the matter axioms will be strictly satisfied. Light is a much simpler physical object than a material rod, and when searching for a relation between the two it should be initially supposed that it would not correspond to so ideal a scheme as the posited matter axioms. Perhaps the matter axioms only have the validity of a first-order approximation in the same way that the ideal gas law cannot be maintained if the accuracy is increased. We can expect a more rigorous agreement solely with clocks, but not mechanical clocks or rotating planets, only from *atomic clocks*, because the frequency of light and the speed of light are related phenomena and could conceivably be placed in a simple relationship. The now clearly emerging red shift of the spectral lines also speaks clearly to this point, however, it belongs to the general theory of relativity and therefore need not be mentioned here (§42 in A.).

What then is proved by the well-known electrodynamic experiments of, e.g., Wilson and Trouton-Noble? It proves that for electrodynamics, the light geometry is much more important than the geometry of rigid rods and clocks. Their explanation does not employ the matter axioms. To the contrary, it is more likely that the laws of electrodynamics are related to the metric for *light* and not the metric of material things. So it appears then that there is a possible development – which has already been started – to eliminate the use of rods and clocks (apart from the atomic clocks) in defining the metric because these things are too complicated and thereby inaccurate mechanisms. The construction of the light geometry that makes the theory of relativity independent of the matter axioms shows that measurement is not eliminated altogether; to the contrary, it provides a more accurate foundation for measurement.

In this way, the Michelson experiment only signifies a single block in the edifice of the theory of relativity and if historically it served as a pillar in the logical system, it merely provides the connection between the light-geometrical basis and the theory of material things. But to what extent this connection is interpreted correctly by the theory only future experiences can teach us.

13

Has the Theory of Relativity Been Refuted?

Since the dispute over the theory of relativity has begun to die down over the last several years and the new theory has been even more successfully worked through, the most recent attacks upon it have come from the flank from which it was the least expected. They are not attacks on the philosophical motivation, and thereby not the well-known reproaches that the theory is "inconceivable" or "incompatible with common senses"; rather, we are now confronted with a physical experiment that stands in explicit contradiction to an assertion of the theory of relativity. This experiment was conducted by the American D. C. Miller at Mount Wilson and was published in the *Proceedings of the National Academy*, Washington (11, 382, 1925).

It concerns the so-called Michelson experiment, one of the most foundational pillars upon which the theory of relativity is constructed. This experiment traces back to the ideas of Maxwell,

Translated from "*Ist die Relativitätstheorie widerlegt?*," *Die Umschau*, vol. 17, no. 30. (1926), pp. 325–8.

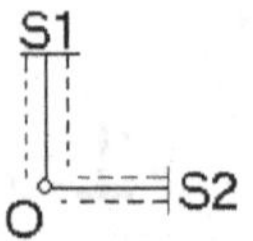

Figure 13.1. Diagram of the Michelson experiment.

but it was Michelson, a scholar famous for his precision in optical measurement, who first carried it out. Michelson had already begun his investigation in the seventies in Berlin as an assistant to Helmholtz, and carried it out in the eighties in America. We can describe the experiment in schematic form in the following way (Fig. 13.1): two rigid arms are placed at right angles with mirrors S_1 and S_2 fixed perpendicularly to the arms at their end points. A light ray will leave the source *O*, travel the length of the arms to the mirror where it will be reflected; what is under investigation is whether both rays having left *O* at the same time also arrive back at the starting point together. According to the old aether theory, this could only be true if the apparatus was at rest with respect to the universal aether; but since the apparatus moves along with the Earth through space, the theory demands a deviation: the ray reaching S_2 must return a little bit later. This experiment, with its very precise measurement involving interference patterns, will thus be able to determine the state of motion of the Earth relative to the aether. In 1883 and again in 1887 with the assistance of Morley, Michelson arrived at his result that stunned the scientific world; in spite of the extreme precision of the measurement, he showed that there is no difference in the time to traverse either arm of the apparatus. In 1904/1905, Morley and Miller replicated this negative result with greater precision.

These facts can only be explained by the aether theory if one assumes that a certain stratum of the aether is dragged along in the Earth's orbit in that same way that a moving ship drags along a thin sheet of water. But this contradicts other facts of observed optical phenomena; for example, we cannot explain why light from the stars is not deflected when passing through this stratum. Indeed, stellar aberration is a well-known fact in which the ostensible location of the star depends upon the relative velocity of the star with respect to the Earth; but the dragged aether theory cannot explain the amount of aberration. So theoretical optics found itself in a seemingly irresolvable dilemma, when H. A. Lorentz in Leyden presented his explanation that assumed that all rigid bodies moving in opposition to the aether undergo a contraction. This contraction only affects the arm OS_2 because it lies in the direction of motion, and it is therefore ascertained that the resulting contraction of the path compensates for the slower speed of the light rays. According to this theory there exists an aether wind over the surface of the Earth, but it is not detectable. Then in 1905, a more basic explanation was proposed by A. Einstein with his theory of relativity in which these contractions occur as a result of a universal principle, the principle of relativity, according to which all systems moving uniformly with respect to the Earth (the so-called Newtonian inertial systems) are equivalent and it is in principle impossible to observe anything that would distinguish any one of them from any other. According to Einstein, this relativity even extends to time; every system has its own sense of simultaneity, and it is meaningless to label any particular definition of simultaneity as the true one amongst all of the others.

This theory has achieved general consensus among physicists and was extended by Einstein in his general theory of relativity in 1915 to also include gravitational effects within its purview. A series of important physical findings were derived by Einstein such as the equivalence of mass and energy, the bending of light in a gravitational field, and the advance of the perihelion of the planet Mercury, for example. Time and time again physical observations have confirmed Einstein's brilliant theory; observations of the red shift of a companion star to Sirius (only observable through a telescope) provided a shining confirmation of Einstein's prediction.[1] Yet, an objection is now raised from an unexpected place which challenges its seemingly unshakable foundation.

D. C. Miller, author of a well-known textbook on acoustics and a widely recognized researcher, who reproduced the negative result of the Michelson experiment with Morley in 1904/1905, has recently conducted the experiment under different conditions. He is an opponent of Einstein's interpretation of the experiment, believing in the dragged aether. He thus conjectured that at some distance from the Earth the dragging would no longer occur completely and thus, at that point, a weak relative motion of the aether with respect to the Earth, a so-called "aether-wind," must exist. To prove this, he placed the apparatus of 1904/05 on top of a tall mountain in order to distance it from the surface of the Earth. After a few preliminary experiments, he chose Mount Wilson, the mountain on which one finds the famous astronomical observatory whose equipment he had access to. In 1921, he found that he could not replicate the negative result. Then, to be sure, he took the apparatus down to the plains of Cleveland and found that there he did reproduce the original negative finding. He took further pains to avoid sources of error; he eliminated all parts made of iron from the apparatus, constructed the arms from brass and concrete, and transported it back up Mount Wilson where, in September 1924 and March/April 1925, he took the decisive measurement. With

[1] For a summary presentation of the new observational evidence for relativity, see G. Joos, *Physikal. Zeitschr.* 27, 1926, p. 1.

American generosity, he was able to make 5000 individual measurements in all and detected the same positive result that was seen in 1921.

What should we now make of these results? It does not mean that we ought to believe that there is a direct demonstration of the aether-wind. In this respect, the negative result is a lot easier to explain than the positive one. The effects that arise from the forward motion of the Earth overlap with those that arise from its rotation about its axis; consequently, the amount of the effect (that is, the time difference for light to travel the different arms of the apparatus) is dependent upon the time of day. Miller represents this dependence graphically where time is plotted on the horizontal axis and the angle of deviation of the direction of the observed aether-wind from the north-south line on the vertical axis. As the calculations show, this must produce a curve similar to the inverted cosine curve intersected by the horizontal axis in the middle (Fig. 13.2).[2] The curve which Miller has found (Fig. 13.2) bears a certain similarity to the theoretical expectation; there is however an essential difference, that the horizontal axis does not intersect it in the center.

Can we then say that the aether-wind hypothesis has been proven? Miller advocates this view with complete certainty and calculates a relative velocity of 10 km/sec, about one third of the orbital velocity of the Earth. But one cannot admit this conclusion with certainty. On the one hand, the curve determined by Miller deviates from the expected symmetry with respect to the horizontal axis. On the other hand, J. Weber has shown that in Miller's crucial figure some quite problematic measurement data has been omitted (the crosses in Fig. 13.2), with no account of

[2] This figure is from J. Weber, *Physikal. Zeitschr.* 27, 1926, vol. 1.

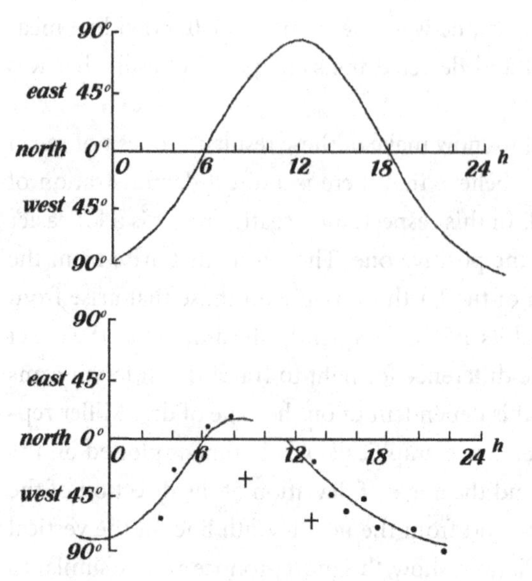

Figure 13.2. Miller's results. Top: calculated; bottom: observed.

why Miller has omitted these points. Indeed, Miller's publication is so short that one cannot possibly consider it a proof of its assertion; Miller would first have to provide the details of his method of calculation so that it may be open to critical scrutiny. When one considers the individual figures that Miller employs in producing his main figure, the agreement appears to be quite unsatisfactory.

Indeed, it must be recognized that there is an inferential leap taking us from this sort of observation to the positing of an aether-wind. Even from the standpoint of aether theory, it is unlikely that there would be a 10 km [per second] aether-wind detected on a 1750-meter-tall mountain given that at the level of the lowlands, the entire aether is dragged along. The height of Mount Wilson is a mere 0.03% of the Earth's radius of 6370 km; that the aether-wind would increase from 0 to one-third of the expected overall amount over such an insignificant interval is improbable. The amount of aether-wind that Miller found is thus too large to seem credible. It must be presupposed that the entire solar system itself is moving with a speed of several hundred kilometers per second through the aether to justify Miller's observations. This is, however, compatible with widely held astronomical hypotheses. Some distant spiral nebulae display this sort of velocity relative to our solar system; if one assumes that these nebulae are giant Milky Way systems then they could be the masses that determine the rest state of the aether. Then it is not only our solar system, but rather the entire Milky Way that moves with this sort of velocity relative to the aether. But, of course, it seems a very unlikely hypothesis that the masses that are combined in our Milky Way should not determine the local rest state of the aether around us.

Hence, before we assent to such radical consequences, let us consider another interpretation of Miller's observations. Perhaps it is a matter of an effect occurring at the height of mountains during certain times of day and that therefore only appears to be an aether-wind. This may lead us to first consider the possibility of the influence of temperature, perhaps connected with thermal radiation; however, Miller has been able to show that this is not possible. Much experimental work should be expected before the true origins of Miller's observation will be explained; and we hope that similar experiments will be conducted by other observers on other mountains.

In this respect, the recent observations of German physicist Tomaschek on Jungfraujoch at a height of 3457 m are significant.[3] Tomaschek has not performed the Michelson experiment,

[3] *Annalen der Physik* 78, 1925, p. 743.

but two others – the so-called Röntgen-Eichenwald and Trouton-Noble experiments – that are important for the theory of relativity, and thereby also for the possibility of detecting an aether-wind. Both experiments yielded the same negative results in both high altitude and the lowlands. When conjoined with Miller's result, we are forced to say that the effects of the aether-wind are detectable in the Michelson experiment, but not in these other foundational experiments. Of course, the careful, thoughtful physicist sees that this possibility cannot be ruled out *a priori*, and we must therefore hope that the Michelson experiment will be replicated on Jungfraujoch. Since Tomaschek had already performed these experiments several years ago in Heidelberg, he would be the most appropriate candidate for their verification, especially since he himself is not counted among the advocates of the theory of relativity and yet nevertheless has proven to be an extremely objective observer even where his results have spoken in favor of relativity theory.

What then does the theory of relativity have to infer from Miller's experiment? Should Miller's results be confirmed, it would indeed be a very strong blow to its foundations. Einstein himself has recently said in the newspapers that if it were found to be the case, he then must declare the principle of the constancy of the speed of light to be false. I have put forward a less radical opinion.[4] The Michelson experiment, of course, played a crucial role in the historical development of the theory; it was an unexplainable anomaly for the old theory of optics that first found clarification through the theory of relativity. But it does not occupy this

[4] See for this and what follows H. Reichenbach, *Axiomatik der relativischen Raum-Zeit-Lehre*, Vieweg, 1924, and also *Zeitschr. f. Physik* **34**, 1925, p. 32 [Chapter 12 in this collection].

same significant place in the relativistic theory's logical structure. Under the ten axioms of the theory of relativity as I have laid them out, i.e., its ten most basic empirical propositions, there is only one that entails the Michelson result; it is only this axiom then that is thereby threatened. The principle of the constancy of the speed of light could be maintained in a more limited form even if the Michelson experiment's negative result were overturned. One could construct a "light geometry" using light signals but employing no rigid rods to maintain a metrical understanding of the world and allow the previous formulation of all physical laws. From this perspective, the Michelson experiment serves only as a bridge between the light geometry and the geometry of rigid rods. Should this connection be lost, this would only mean that rigid rods do not after all possess the preferred properties that Einstein still attributes to them. This would not mean a return to the old aether theory, but rather a step towards the renunciation of a preferred system of measurement in nature. Before accepting such a move, Miller's result must receive some very credible confirmation. Einstein has also clearly explained in his above-mentioned remark that he cannot take Miller's result as compelling at this point.

Miller's result in no way affects the philosophical consequences of the theory of relativity. The investigations of the last few years into the logical foundations of the theory of relativity have clearly exposed what are the philosophical underpinnings and what are the physical results of the theory. It thus has been stated with complete rigor that the actual physical assertions of the theory are not derived from philosophical propositions, but that they merely represent one possible way to flesh out the general framework that must accommodate the empirical results. From this point of view, the framework is independent of specific, physical observations.

In the epistemological foundations of the theory of relativity we have the logical results before us that flow from a particular physical theory, but which have also given rise to philosophical insights which no longer belong to the realm of physics but rather to the philosophy of nature.

14

Response to a Publication of Mr. Hj. Mellin

1. Mathematics and reality. 2. Time order. 3. Simultaneity. 4. Uniformity.

Recently, Hj. Mellin, in a lengthy examination,[1] offered a critique of my *Axiomatization of the Theory of Relativity*.[2] A discussion of the objections that Mellin raises to my axiomatization, and thereby to the theory of relativity, seem to me to serve the general interest because of their fundamental nature and his clear formulation of views which are most often only operative on a subconscious level, and I would therefore like to answer them here.

1. The most significant difference in our positions lies in our understandings of the relationship between the mathematical

Translated from "*Erwiderung auf eine Veröffentlichung von Herrn Hj. Mellin,*" *Zeitschrift für Physik*, vol. 36, nos. 2–3 (1926), pp. 106–12.

[1] Hj. Mellin, "Kritik der Einsteinschen Theorie an der Hand von Reichenbachs Axiomatik der relativischen Raum-Zeit-Lehre," *Ann. Acad. Scientiarum Fennicae* (A) **26**, Helsingfors, 1926.

[2] H. Reichenbach, *Axiomatik der relativischen Raum-Zeit-Lehre*, Braunschweig, Fredr. Vieweg & Sohn Akt.-Ges., 1924. See also *ZS. f. Phys.* **34**, 32, 1925 [Chapter 12 in this volume].

discipline of geometry and reality. Here, I adopt the perspective (which is often incorrectly termed conventionalism) that the geometrical axioms as mathematical propositions are not at all descriptive of reality; this only occurs when physical things are shown to be coordinated to the elements of geometry (coordinative definitions). If we take very small bits of mass to be points, light rays to be straight lines, and the length of a segment to be determined by the repeated placement of a rigid rod, then the statement that straight lines are the shortest becomes an (empirically proven) statement about real things. Without such coordinative definitions, these propositions say nothing about reality.

Mellin's objections to this view rest on his stressing the so-called intuitive necessity of the geometric axioms. Accordingly a geometric axiom cannot be arbitrarily posited; rather, there is a compulsion of the intuition which drives us to the axioms. It appears to me that this view – which of course I hold to be incorrect – is irrelevant to the problem at hand. For even if logic alone is not considered decisive for the determination of mathematical truth, and some intuitive criterion is also required, still nothing follows about the relationship between the axioms and reality; this intuition must certainly not be confused with perception; rather, the "pure intuition" must be separated from the "empirical intuition."

Thus the stated objection cannot demonstrate the epistemological necessity of the coordinative definitions. And that these definitions are arbitrary follows directly from the fact that the coordination between elements of the geometry and real things is not prescribed. Hence, there is no criterion of truth for definitions as Mellin would like to require using perception (p. 7). Physical propositions can only be true or false if they are based upon coordinative definitions. For example, we might say that the above sentence is false when the concept straight line is not coordinated to the path of a light ray, but rather to the path a boomerang traverses through the air; but we cannot call this coordination false. One could maintain this coordination and then redefine the metric so that straight lines are again the shortest. We are not forced to accept any given coordination between concepts and objects; first one requires that known propositions (e.g., the axioms of Euclidean geometry or non-Euclidean geometry) are said to be true, and then when some concepts are coordinated to real things, constraints arise for further coordinations.

This holds for the basic concepts of the topology of space as well. Mellin cites the concept "between" as an example and claims that only the intuition can determine which of three points *A*, *B*, *C* lies between the others (p. 8). But he forgets that – even if the intuitive character of the concept of betweenness is admitted, which I do not – it means nothing to reality unless one confuses intuition with perception. If we say that we make an observation of point *B* lying between *A* and *C*, it does not yet follow that *B* is really between *A* and *C* (think perhaps of an illusion of perspective); but in order to make this determination in another way, we need a specific description of what it means to be physically between, hence a coordinative definition. In my axiomatization this coordinative definition is not specifically given, but is implicit in the general definition 11. An explicit definition, though, could be given in the following way:

Definition e. A point *B* lies between *A* and *C* if the first signal traversing *ABC* arrives at *C* at the same time as the first signal traversing *AC* (abbreviated $\overline{ABC} = \overline{AC}$.)

For this definition to agree with the pure mathematical meaning of the concept "between," we must add the empirical statement:

Axiom G. For two given points B_1 and B_2 such that $\overline{AB_1C} = \overline{AC}$ and $\overline{AB_2C} = \overline{AC}$, then either $\overline{AB_2B_1} = \overline{AB_1}$ or $\overline{B_1B_2C} = \overline{B_1C}$.

This determination of the concept between then directly leads to the definition of straight line as the "field of betweenness relations":

Definition f. The straight line through A and C is the set of all points that satisfy the betweenness relation among themselves and include A and C.[3]

The statement that the straight line is the shortest will then be a provable proposition based upon the other axiom. This example clearly shows that independent of all purely mathematical definitions specific coordinative definitions are necessary when one wants to apply the geometrical concepts to reality; and at the same time it shows that this applicability again assumes particular properties of reality.

Since my axiomatization takes this route, gives coordinative definitions of geometrical concepts, including time, and formulates their applicability in the axioms, I do not understand how one can raise the objection that this procedure would "construct reality from pure thought" (p. 4). To the contrary, geometry is constructed directly from real things and not impregnated with pure thought; indeed the coordinative definitions are there for this

[3] This definition is more correct than definition 9 used in my axiomatization.

purpose. Mellin seems to especially doubt the real meaning of the light geometry; but it does not matter whether the coordinative definitions relate to light or material measuring devices, since light is every bit as much a real thing as more solid pieces of matter. The light geometry is just as real a geometry as material geometry. In my axiomatization, I use the term "axiom" to make clear which statements are empirical. When Mellin says, "Stressing observation as the ultimate criterion of truth in physics loses all meaning because the observations have no influence on the axiomatization" (p. 17), it is nothing but a misunderstanding. My axioms are not "intentionally constructed and interpreted to the demands of (Einstein's) theory" (p. 17), but rather are nothing but the uninterpreted observation reports which lie at the foundations of Einstein's theory and can be confirmed independently of it. Let me remind him of such axioms: axioms IV, 2, and VIII have been literally proven through experimentation. When I have referred to the axioms as steps towards greater abstraction, I mean it only in the sense that this is how it is for every so-called physical fact. It is beyond dispute that we can never have direct proof of any physical proposition; rather, we move from perception to observation through the mediation of theory; it is no different for the so-called "directly provable statements," although the content of the theory employed in these cases plays a much smaller role. But this does not relieve perception from its position as the ultimate criterion of truth.

While the discussion of the proposed objections concerns a fundamental question and while I have only given a sketch of my response, I can go on no further as I must now move on to deal with particular objections in which Mellin contends my axiomatization proves inconsistent.

2. First, Mellin holds my definition of time order is full of contradictions. He says that "the concept of signal as a physical process presupposes time and therefore also the concept earlier/later" so that I thereby cannot turn around and define time order in relation to signals. This, however, is obviously mistaken. I explicitly use a process that allows the determination of the order of cause and effect without referring to time order, and then define time order in terms of it. This process makes use of the basic proposition that there is a unique causal direction in which events can leave a mark; I produce the mark of the "cause," and I find it reproduced in the "effect," but when I produce the mark of the "effect," it does not appear on the cause. To use a term different from that employed in my axiomatization, consider the following:

Let Q_1 and Q_2 be two events which we know to be causally related, but we do not yet know in which direction. I now conduct repeated experiments first marking one and then the other event and call the one with the mark event Q'. For the moment, let us represent the causal relation with a two-headed arrow indicating our temporary uncertainty of the direction; we then write the results of our observations in the form:

$$\text{We can establish: } \begin{cases} Q_1 \leftrightarrow Q_2, \\ Q'_1 \leftrightarrow Q'_2, \text{ but not} Q'_1 \leftrightarrow Q_2. \\ Q_1 \leftrightarrow Q'_2, \end{cases}$$

In these relationships, Q_1 and Q_2 are asymmetrical, and we can therefore define the direction of the relation so that we can obviously draw the arrow such that:

$$Q_1 \rightarrow Q_2.$$

In other words: we consider the event that stands in the relationship without prime to be the effect, which also happens to be later in time.

This is probably sufficiently clear to show that temporal consciousness is unnecessary for determining the order of Q_1 and Q_2 as Mellin believes[4] (p. 20) and that my definition of time order is in no way logically circular.

3. Mellin further objects to my definition of simultaneity. He accuses me of being circular here as well and claims that my proof of proposition 6 on page 26 assumes absolute simultaneity because the phrase "we synchronize all clocks with a central clock" occurs there. He overlooks, however, that this synchronization, defined in definition 2, is arbitrary within the limits of this definition and requires no further assumptions within which we would find absolute time. There is no circular reasoning here either.

At a later point, Mellin puts forth a method to develop an "epistemology" of absolute simultaneity. Let AB and $A'B'$ be two rulers of the same rest length that are sliding alongside each other. According to Mellin, we can "establish through many visual and tactile experiments that as soon as A and A' coincide, that at that moment B and B' simultaneously coincide" (p. 36). This determination is flawed. There is not a single experiment by which we could establish this. The so-called direct perception fools us with its lack of acuteness because it neglects the causal chain that runs from coincidences to our sense organ; but the arbitrariness

[4] Mellin contends (pp. 19–20) that there is a further error in my axiomatic presentation when I call *S*(*P*) (in my original terminology) an event. However, in axiom I,3 I have explicitly posited the possibility that every point in time can be considered an event even if there is no coincidence with a light signal.

of relativistic simultaneity in fact lies only in time intervals of this order. Any exact measurement has to involve an assumption concerning simultaneity at distant points. This is obviously a question only of the definition for simultaneity; and as such, if the axioms of the theory of relativity are correct, it is not unique, i.e., at different speeds the moving rulers will not produce the same simultaneity. Besides, I have already dealt with Mellin's definition of simultaneity on p. 100 of my axiomatization where I show it to be utterly useless; it had previously been suggested by H. Grunsky.

4. It is curious that at one point in Mellin's position, he acknowledges the necessity of a coordinative definition, if only indirectly. In particular, he disputes the existence of an absolute uniformity of time and says, "no motion is intrinsically uniform, but only in relation to something else is it either uniform or not" (p. 40). This is exactly what the theory of relativity contends when it constructs coordinative definitions. Our basic tenets applied to the notion of uniformity yields the following: since we are given no absolute uniformity, we must arbitrarily select some state of motion as uniform and use it to fix a measure of time as a relation of successive temporal intervals. We do this to fix the scale just as we do in the case of temperature; however, we do not assert that we have thereby chosen a state of motion that determines "true uniformity," but we explicitly hold that every choice is equally legitimate. The statement that a coordinative definition is necessary for a concept is therefore completely equivalent to the statement that a physical object can only be measured in relation to another physical object and not with respect to some pre-existing measurement in itself.

The theory of coordinative definitions therefore in no way serves as a "tower of Babel," but to the contrary creates strict order in being able to point out that physical propositions only have meanings as relationships between two or more physical objects. Further, Mellin errs when he believes that the relativistic concept of uniformity is "accepted untested" by the theory of relativity (p. 41); to the contrary, my definition 12 provides a coordinative definition of uniformity just as it does for other metrical notions.

5. With this response, I have defused the heart of Mellin's objections, so it appears unnecessary to address any further details.

Index